Library of
Davidson College

MORE STUDIES IN EARLY PETROLEUM HISTORY
1860—1880

A volume in the Hyperion reprint series
THE HISTORY AND POLITICS OF OIL

HYPERION PRESS, INC.
Westport, Connecticut

MORE STUDIES IN EARLY PETROLEUM HISTORY

1860 — 1880

BY

R. J. FORBES

WITH 2 TABLES AND 36 FIGURES

LEIDEN
E. J. BRILL
1959

338.4
F694m

Published in 1959 by E.J. Brill, Leiden, Netherlands
Copyright 1959 by E.J. Brill, Leiden, Netherlands
Hyperion reprint edition 1976
Library of Congress Catalog Number 75-6473
ISBN 0-88355-291-4
Printed in the United States of America

Library of Congress Cataloging in Publication Data

Forbes, Robert James.
  More studies in early petroleum history, 1860-1880.

  (The History and politics of oil)
  Reprint of the ed. published by E.J. Brill, Leiden, Netherlands.
  Bibliography: p.
  Includes index.
  1. Petroleum — History. I. Title. II. Series.
[TN870.F728   1976]   338.4'7'665509   75-6473
ISBN 0-88355-291-4

CONTENTS

	page
Preface	VII
I. Pierre Belon and Petroleum	1
II. Matthiolus and Petroleum	16
III. The Chemists and the Composition of Petroleum	37
IV. Naphtha goes to War	70
V. Oil from Eastern Europe	91
VI. Oil for Millions of Lamps	108
VII. Wax for our Candles	144
VIII. New By-Products for Burners, Bearings and Bitumens	159
Index	193

"It is a blessing to be given opportunity to improve ourselves by taking warning from the mistakes of others, and in all chances and changes of this mortal life to be free to copy the successes of the past instead of being compelled to make a painful trial of the present."

Diodorus, Library of Universal History
Book I, chapter 2

PREFACE

This second series of essays in early petroleum history is dedicated to the Fifth World Petroleum Congress to be held at New York in 1959 during the centenary of the first succesful completion of an oil well at Titusville, Pennsylvania by "Colonel" Drake. This book and the previous volume were inspired by the succesful sessions of the historical section of the Second World Petroleum Congress (Paris, 1937) and we sincerely hope that a historical section will again be included in future Petroleum Congresses now that so many oil companies have sponsored the publication of historical documents illustrating the growth and scope of their business. Most of these essays are devoted to a description of the little-known European petroleum industry which flourished in the times of Drake and which rose against heavy competition of the cannel-coal, lignite, peat and tar-industries supported by research on petroleum carried out in the university laboratories of Europe. This branch of the petroleum industry deserves our attention for in this competitive atmosphere it laid down the basic operations for the production of petroleum products other than lampoil, such as fuel oil, paraffin wax, gasoline and asphaltic bitumen and found applications for products which were not in common use in the United States for many a year to come. This development is shown by extracts for the original documents in order to carry the reader back to the days when the foundations of our industry were laid.

Maurice Mercier was kind enough to allow us to publish an English version of the chronological table on Greek Fire and gunpowder, which he published in his excellent book "Le Feu Grégeois" (Geuthner, Paris, 1952).

December 1958   R. J. Forbes

CHAPTER ONE

PIERRE BELON AND PETROLEUM

When discussing views of the sixteenth century on the nature and origin of petroleum we usually cite [1] Georgius Agricola's detailed handbook on mining and metallurgy, in the twelfth book of which he deals with the production and refining of crude oil and bitumen [2]. However, Georg Bauer (1494-1555), thorough and accurate as his descriptions may be, was familiar only with the seepages and crude oils of Central Europe. A far more complete survey of the crude oils known at that period can be found in one of the writings of Pierre Belon (1517-1564), which is seldom consulted as Belon is of course in the first place a botanist and a comparative anatomist of world fame. However, during his travels in Central and Southern Europe and the Near East he personally visited many seepages, which his contemporaries knew from hearsay only or from passages in classical authors. Hence it is worth while to study Belon's survey published a few years before Agricola's account.

We are fairly well informed about Belon's life [3]. He was born at La Soultière in 1517 and entered the service of the bishop of Clermont at an early age. We then find him among the followers of René du Bellay, bishop of Le Mans. In these typical Renaissance surroundings, characterized by a profound interest in the arts and in nature, his interest in botany and zoology was awakened; and in 1540-1541 his master sent him to Germany to study at Wittenberg. Here he met his contemporary Valerius Cordus, famous botanist and pharmacist (1515-1544), whose lectures he attended and whom he joined with several fellow-students on a botanical excursion which took them through the whole of Central Europe. After a short trip to England we find him studying at Paris, where he attracted the attention of Cardinal de Tournon and other influential courtiers. His great intellectual powers made his protector

[1] R. J. Forbes, Studies in Early Petroleum History (Brill, Leiden, 1958, Chapter III).
[2] Georgius Agricola, De Re Metallica (transl. H. C. Hoover and L. H. Hoover; Dover Publications, New York, 1950); De Nature Fossilium (Basel, 1546, Book IV).
[3] P. Delaunay, L'aventureuse existence de Pierre Belon du Mans (Revue du Seizième Siècle, vol. IX, 1922, pp. 251-268; vol. X, 1923, pp. 1-34, 125-147; vol. XI, 1924, pp. 30-48, 222-232; vol. XII, 1925, pp. 78-97, 256-282).

entrust him with diplomatic missions to Switzerland (1542) and Germany (1543). On his way back home he was imprisoned for six months at Geneva, but he was freed after many negotiations.

With his life-long friend Cordus he now travelled to Provence and Italy, collecting many new botanical species. Back at the French court of Francis I he propagated the introduction of many new useful plants into France. The king, with whom he stood in high favour, sent him with the French Ambassador Extraordinary, Monsieur d'Aramont, to Constantinople and thus gave him a chance of exploring the Near East. He left Venice in December 1546 and reached the Turkish capital by way of Corfu, Zante and Crete. Once at Constantinople he set out on an excursion to Macedonia and Thrace whence he returned to his base in August 1547. Together with M. Frumel he then travelled to Egypt and returned overland by way of the Sinai desert, Jerusalem, Damascus, Syria, Cilicia and Armenia, arriving at Constantinople in 1548 and returning home to France during the next year, not forgetting to explore the Adriatic coast on his way back. In 1549 and 1550 he was at Rome with Cardinal de Tournon at the conclave held there, which visit gave him a chance for further explorations in Italy. After his return to Paris he crossed over to England twice. King Henry II, impressed by his frequent pleas for acclimatizing useful foreign plants in France, sent him on a trip to Switzerland (1557) and to Italy (1558).

At this stage of his life Belon started to study medicine at the universities of Paris and Montpellier, obtaining his degree on May 18, 1560. Political troubles in the ensuing years 1561 and 1562 robbed him of his most influential protectors at the French court, but he remained in high favour with the king and was allowed to retire to the "castle of Madrid" in the Bois de Boulogne, devoting his time to research and writing as "escolier du roi". In the Bois de Boulogne he was murdered by robbers in 1564.

His main works were published during the years after his return from the East, viz. his Natural History of Fishes (1551), his Observations on Several Singular and Memorable Things found in Greece, Asia, etc. (1553), his De Aquatilibus (1553) and his Natural History of Birds (1555). The work which merits our particular attention [1] was written in Latin. It lists archaeological memories of his travels in Egypt and the Near East and deals more particularly with the burial customs of the ancients, the arrangements of tombs and sepulchres and the substances used to preserve the dead bodies. He obtained

---

[1] Pierre Belon, De admirabili operum antiquorum et rerum suspiciendarum praestentia (B. Prévost, Paris, 1553) (The chapters on petroleum are found in Lib. III, cap. V-VII, pp. 43-49).

official permission to print it on June 21, 1553, and it was published on July 8, 1553.

In the third book of this De Admirabili he points out (page 34 verso) that one of the favourite drugs of his time, mûmîa or mummy, was not identical with the pissasphaltos or Jew's pitch of the Dead Sea, as so many contemporaries like Matthiolus believed on the strength of the testimony of ancient classical pharmacists like Dioscorides. Belon gives details on the manner in which this mummy is collected from the scrapings of mummy bandages and the mixtures of herbs and resins with which the bodies were filled after the intestines had been removed. He also points out that "artificial mummy" was being manufactured by exposing dead bodies to the heat of the desert sun and excavating them after some time to grind them up into the famous "mummy". He therefore declines to use mummy for pharmaceutical prescriptions, though only some of his fellow scientists agreed with his views, others persisting in the use of this valueless and even dangerous drug [1]. We will not go into details on this problem but rather turn our attention to the chapters on petroleum (or naphtha) which Belon wrote after having discussed mummy (or so-called pissasphaltos) and the natural bituminous substance which according to him is rightly called pissasphaltos by the ancients.

This passage of De Admirabili .... reads as follows:

"CONCERNING RED NAPHTHA, OR RED GLUE OF BITUMEN
by Pierre Belon of Le Mans
Chapter V

But now that we have sufficiently discussed bitumen (asphaltos), or Jew's pitch (pissasphaltos), it remains for us to discourse of the glue of bitumen, for when we began to write of those things in particular that are used for preserving the body it seemed to me that Naphtha, being a substance related to bitumen, is worthy to be described.

Among the Turks and Arabs Naphtha is still known as Nepht at the present day; but we, having forgotten its ancient name, call it Petroleum. However diligently I have sought out its origin, and not without a great deal of labour, it certainly seems to be useful to note down the several things that appear relevant to its history. It is wrongly believed by some that naphtha is nowhere found in Italy, because the latter never seems to have been mentioned by the

---

[1] R. J. Forbes, Studies in Early Petroleum History I (Brill, Leiden, 1958, chapter XII).

Arabs and Greeks but only attributed to the Babylonians. It is very likely that naphtha was found in the region around Babylon since Asphalt and Jew's Pitch are obtained thence. Naphtha, which is abundant in Italy, burns if fire is applied to it; in its source, however, it does not burn in the way it does in Babylon as was noticed by Plutarch in the Life of Alexander the Great. For he writes that when this king had traversed and conquered the whole of Babylon which had straightway submitted to his rule, nothing astonished him so much as a crevice in Ecbatana (Hamadan) from which fire issued continuously as from a fountain, and naphtha flowing out of it, such an abundance being belched forth not far from the crevice as to form a pool of which the proper nature is to attract fires to itself. In his chapter on Asphalt Serapion has also written much on the subject of Naphtha which would be too long to recount here. But we thought it worthwhile to give a summary as far as concerns the entire account of Naphtha, which we saw in abundance in Italy. Dioscorides rightly terms Naphtha "glue of bitumen". Indeed nothing has ever been found to retain the odour of Naphtha better than Bitumen. In the very place where Naphtha issues it smells of Bitumen in a wonderful manner. There are three chief kinds of Naphtha in Italy, *viz*., white, black and red, and these are still abundant in various places. There is a spring beneath Mount Celizi from which they obtain Red Naphtha. There is another spring that issues at the village of Meiano (Miano de Medesano?) whence white Naphtha is derived. There is also a third spring from which black Naphtha issues in greater abundance in the village of Allensi. We decided to write of these places separately so that if any other student of nature and history should wish to see its spring he might at least know whether to go. The inhabitants mix the three oils together so that it is unusual for an unmixed oil to be transported to us. But the naphtha oil that gushes out near Mount Zibito (Zibio) 13 miles from Modena is sold at a higher price than that obtained from Salsa. In fact, the oil won at Salsa, as I said, is black and muddy and lower in price. This is why the two are mixed together and we are thus defrauded. Moreover neither is transported to us by itself in an unmixed form, but only the three, black, white and red, mixed together. But this is not a subterfuge of the local inhabitants, but of merchants who when they purchase three kinds from many individuals, they afterwards mix them together and send them abroad thus mixed. The red naphtha obtained from Mount Zibito is most highly esteemed. This is because it exceeds the others in price and is also found to be greatly superior in other qualities. The nature of its tract or region is such it sticks very tenaciously to the feet of wayfarers as though it were dripping wet. The soil is gravel, of a somewhat whitish colour suitable for making tiles, as it may be worked up into all kinds of earthenware. I have been informed that this red oil is obtained from a cave. Nothing is known for certain as to when it was first discovered. Some say that this oil was first excavated from an underground passage 150 years ago. The natives believe it was a pig that first came across this oil and give the following reason for their belief. When the pig constantly wallowed in a morass on which oil was floating and carried back home the stench or odour of the oil, the servants wondered what it might be that smelt so strongly and realized at length that the smell came from the pig. They therefore went to the morass and seeing the oil floating thereon they carefully collected it, and first they used it for fire, employing it in a lamp-wick during their night watches, but afterwards decided to send it abroad on account of its fragant odour, believing it would greatly increase in value in the future on account of its marvellous nature (because it was so greedy of fire). There are two or three wells of this red oil in the valley below Mount Zibito, six miles from the village called Saxolo (Sassuolo). Should you wish to see this oil, starting from Modena you go to the village called Saxolo or Sasolo (Sassuolo), and then go up the mountain. On the summit of Mount Celizi there is usually a slight shelter. One then descends for a mile towards the foot of the mountain in order to see the cave I refer to. Beside the cave a stream flows out of the hill through a deep valley into the plain. No single person owns or controls this oil, nor is there an inhabitant who lays greater claim to it than any other. Originally there was a dispute among the inhabitants about this spring; the citizens of Modena referred to the judges their dispute with the inhabitants of a place in the neighbourhood called Revina, which is opposite that part of the mountain facing west.

At length, the poorest peasant of all, who was entitled by law to the estate where oil flowed, since he could not break the opposition of so many, with his friends and relations who at that time numbered only thirty, he staked his claim, and in the end they won the case against them. At first the oil flowed in conjunction with water from the ground, and from it the inhabitants could derive nothing or at any rate little without considerable effort. Consequently they excavated the soil until they penetrated 12 feet below ground. When, however, they separated the oil from the water before trading it, and discovered that the well descended into a trench which first belched it up, they refrained from excavating any further. But the spring of this red naphtha well originates from very stoney ground, concealing a small pit excavated in it which leads down to the bottom of a cave. There are two channels connected with it, from the one which is on the right water flows slowly; from the other, which is on the left, oil pours drop by drop. And the naphtha is then decanted every day, or at least every other day. But after they have decanted the oil, they then draw off the water: since if they did not remove it thence, the water would fill the whole of the cave, and the hole through which it passes being thus closed, the oil would no longer trickle down. The cave is a single one, and is closed by a very light door, a long way away from the inhabitants of this place, for they come down from the shelter, a mile away, in order to take away the oil. The key to the door is kept by the peasant to whom is entrusted all the work to be carried out. This oil trickles down gradually at a rate not exceeding a weight of thirty ounces per day, which is equal to the measure of one Bocal, but it does not reach the same figure at all seasons of the year. If there is a better summer, when the mountain becomes heated by the rays of the sun, the trickling increases, and then a bocal and a half flows out. In winter, however, it only yields an average of one bocal, so that by calculating the summer and winter rates, we find that it yields one bocal per day. One can descend inside the cave by stairs or steps. There is a vault in a rather light and modest building, surrounded on all sides by walls, to prevent the access cave from being closed up by the snows, and by the fertile earth, gradually loosened by the rain, and sliding down from the mountains without. The water which trickles down with the oil is clear and limpid, and slightly salt. Its salt taste proves to be much milder than that of the water which I tried at the springs at Larissa, in the Troad. In these hills there are also other springs of this oil, which yield a limited quantity of oil in summer-time, but in winter do not even yield that much. Apparently this oil is nowhere flowing out by itself, but floating on the surface of water. Those who are desirous of collecting it stir the water with twigs and drive the oil into one of the corners, and then draw it off with shells or gourds. There are some among the inhabitants who have digged slanting wells as far down as 60 feet, in which at the time they were very successful. But this was easy for them, since they did not encounter any hard stone in it, but only soft tufa, so that I wonder why the people should call this oil petroleum. I should think, however, that the name Petroleum (stone-oil) was derived by the Italians from an earlier stone shaft. In short, I declare that this oil, wherever it may occur, never flows without salt water. The above-mentioned stone shaft is situated in the deep valley of a steep mountain facing entirely south, between two hills, which is called Mount Saint Peter (Monte S. Pietro). But the other spring faces north, and is called the spring of San Marino. The first spring is moreover called "the ancient spring" on account of its age, to distinguish it from the other, from which black naphtha gushes, and which for this reason is called "the black one". It stands alone in the deep valley near the foot of that mountain which is called Revina, and the quantity of oil that seeps through in one day is not more than six ounces, which flows drop by drop with water from the said mountain, and smells more of asphalt than the red variety, nor is it shut in by any kind of door, and if anybody from Bologna wishes to go to these springs, he should take the road which leads to Confortis (?), which is seven miles from Bologna, then on to Spilamberto (seven miles), then to Maravellae (Castello di Serravalle) (twelve miles), a village which is not more than a mile away from this spring, making thirty miles in all. But although in several springs all three different kinds of naphtha, white, black and red, are to be found nowadays, the ancients only knew the red and black types. The red type, which flows from the ancients spring, I have already described above. It is now time to speak of the black type. I have said that the red and the white varieties are mixed together by impostors, and that the Modena variety, which takes pride of place, is collected near Mount Zibito; however, the black oil which is gathered, which I shall now discuss, is customarily used by merchants to adulterate the white and red types, so that none is taken away unmixed. The source of black naphtha is in that place which in Latin can be called Halis. In Italy it is popularly known as Salsa, because salt is dried there. In that place the ground has to be dug down to a depth of 720 feet before reaching the salt-water spring. However, the reason for sinking the wells to this depth was not only for the sake of the naphtha, but also of the salt. That oil alone is too low in price to suffice by itself to cover the expense involved. But because the water is drawn from the wells in order to extract the salt from it, there are therefore many wells; however, one is more important than the rest, which usually produces a larger quantity of salt water than is yielded by all the rest. Consequently, much less black naphtha is drawn from this one than from all the rest. In winter the same quantity is yielded as in summer, provided the water is deep down at the bottom of the well. For when no more water is drawn as when wood is lacking, or as sometimes happens in harvest-time (when the salt is not extracted), the water rises over the top of the well, and only flows from the spring, forming a stream, and then no oil comes out. That village where the salt-mines are, the Milanese call "selz", pronouncing the letter "a" as "e", when it ought to be called Salsa: for its name has been corrupted to Sale. In fact, the Milanese change a into e in all Italian words. The water in the wells in this village is much salter than sea-water: for when it touches the tongue, it has a more pronounced saltiness than sea-water, and stings the palate on account of its great bitterness. Therefore, distributing the water to the individual houses, they dry out the salt to their great profit. Because, if by chance they do not draw off the water, it flows down into the nearby river, but then it does not take any of the oil with it. But this is the way they draw the water. A large wheel is turned by two men walking inside it, which raises a bucket full of water from the bottom to the top, and carries it up as far as a circular table or floor, where a workman sits and pours the water out of the bucket into a big jar, and when the latter is full, the oil floats to the top: this is collected with a vessel before they send the water through pipe-lines and channels and thence into the iron cauldrons standing in the houses. The cauldrons in which the salt is dried are 1½ feet high and 42 feet long, and being broad they are suspended by five iron rods from the floor table; for they are not supported at the bottom on props, in order that the fire should penetrate better into the salt. Because if it happens that the cauldrons are not properly connected, they stop up the leaks, applying the only remedy to both the elements, namely fire and water. They take as many eggs as they need, in which they put finely-ground flour, and soot lamp black, that is to say, dust blown through the fire-place or only adhering to the walls and scraped from the cauldron, and as soon as they were mixed together, they dip tow in it, and if there are any holes in these cauldrons this (lute) stops them up. Such is the force of this water, that an egg cast into it does not sink. The salt here is never as a rule made up into lumps, as is generally the case in many places in Germany. But the flavour of the salt is more bitter than sea-salt, and therefore it is known that it requires no further drying, while in many places other kinds of salt are left for a long time to dry out in the heat on the table. In this village there is a number of wells, of which the one which discharges Naphtha with great force is said to be about 1,200 feet deep. The method of drawing off the oil is not the same for all of them. Some people dip sheep-skins in it, and soak them in the oil floating on top, and afterwards wring them out. All the oil in this village is black. It may be not inappropriate to describe here the method of digging the wells, as it is easier to ascertain than characteristics of black Naphtha. First of all they dig down into the ground to a depth of 120 feet; at length, after they have reached the soft stone which is popularly known as Giastro or Chiastro, and which I believed to be gravel, they then found that the salt water gushes forth. If we can believe Pliny, this soil could not be gravel, because according to him there cannot be salt water underlying it, only fresh water. For in book 31, chapter 3, he says: "The water is always fresh in clay soil, and comparatively cold in tufa. Gravel only occasionally contains water-courses but with a sweet taste". Moreover, I should not think he is the kind of person to call things by the wrong name. Consequently, I should not lightly dare to state whether or no it be gravel. They excavate this type of stone with fairly light hoes: and the deeper they delve, the stonier they find it becomes, where at last the black Naphtha flows. All the wells in this place smell of bitumen, and the more water is drawn from them, the more oil is obtained. From the large one called the ancient well, I say that the number of bocals yielded every day is sometimes greater, sometimes less. Moreover, it sometimes happens that in summer eight bocals are collected.

However, it is not possible to determine its weight with certainty: for in hot, very dry summers, the oil flows abundantly and is drawn off, but in winter little or none flows out. Such is the affinity of fire for this oil that no light can be taken down the well. I can give an example of this fact which occurred thirty years ago and is remembered by all the inhabitants of this village. A certain man had promised to report back to his patron whence the said spring of oil rises, *i.e.* from South to North or from East to West. This man, being about to lower into the well a lamp, protected it well all the way round, and securely sealed it. And so he ordered that he should be lowered by his servants into the well by means of the wheel which brings up the water. Another man went down before him to the bottom of the well without any fire, to await his companion who was to bring the lamp. But however, when the one who was coming down carrying the lamp and the fire, arrived a little above him, the force of this naphtha oil caused its "fire" to explode. Moreover, a flame suddenly ignited throughout the well, burning the man who was being lowered in the bucket with great force and with a dreadful sound, and he could only utter sounds like rifle-shots which rose up into the air. Moreover, so great was the noise that anyone would readily have thought that it was no other than a bullet shot from a rifle. And such was the blast that the circular floor at the top which covered the well, was lifted up into the air: and as a result of this, the people throughout that village trembled, saying that destruction threatened them, and thinking that their last hour had come. Moreover, they were seized with such fear that they fled, abandoning all their possessions. Moreover, the man who went down first into the water was almost burnt, mostly in those parts of his body which were not under the water; however, he was not killed, but the four people standing by the table lowering his companion, were all killed. However, the said fire was extinguished instantly, and it did not last longer than a flame kindled with fire-drills. Various things are related concerning the naphtha flames, and even in Pliny we find chapters on Naphtha, which closely agree with what Herodotus states. The latter says that in Cissia (near Susa) there is a well which contains three different substances. For from it they draw asphalt (that is to say bitumen), salt and oil, in the following manner: it is actually drawn with a celoneus, that is to say, a bailer; instead of the bucket, however, a wine-skin, divided in two, was tied to the arm of the lever: when this is thrown off into the well, anything which is drawn is immediately poured into a tank, and when it is poured out into another container, it is found to consist of three kinds of substance, always including asphalt and salt. This oil, moreover, the Persians call "Rhadinace". So much for the black kind.

CONCERNING WHITE NAPHTHA

Chapter VII

(the numbering is wrong!)

There is a white naphtha spring at Meiano (Miano de Medesano?), of which there are three sources, a short distance away from the houses in the village, on the way to Parma; however, from these three springs, the oil collected is not more than ½ lb. or 8 ozs. per day. For it flows out in a slight trickle, with a somewhat larger quantity of water, and yielding less oil than that which flows near Mount Celesi or near Salsa. Therefore the inhabitants who are responsible for the work devised by systematic investigation a method whereby the water passed through a channel to the lowest of the outlets, in sufficient quantity so that the oil always floated and none of the oil was lost, so that the place where the water collects should always remain full, and oil can be collected every day or every other day. The same method is used for drawing off the oil at Meiano as at Modena. The water is stirred with twigs, and they drive all the oil into a corner, take it up with a vessel and store it away. But just as this oil is whiter than the rest, it also has more special virtues, and is in fact found to have a better smell. Meiano is 10 miles away from Parma and 10 miles from Salsa, and is situated between the other villages, namely Alausano and Costa Meiano. It is subject to the overlordship of the ruling counts of Froliorossi. The price of the white oil from Meiano is no higher than that of the oil obtained from Modena: but it is no wise inferior as regards its other characteristics. It springs in a thin trickle mixed with water from the summit of the mountain. In the summer it trickles down fairly abundantly, more in dry weather than in rainy weather, and more when the sky is clear than when it is overcast. All this likewise applies to Modena. The price of each does not exceed five Italian solidi per pound, which converted into French currency works out at two Caroli guilders, *i.e.* twenty Tours florins. With regard to the three kinds of naphtha as found in Italy, these are different: to a fairly keen observer, they do admittedly differ in the way they taste, smell and burn, but the intensive difference will not be found to be so great. The black kind is more viscous than the red kind: and the red kind has a slightly higher viscosity than the white, wherefore the red is distinguished only by name from the white: both of which are so fluid that they are not unlike water. If you should see a naphtha which looks slightly darker, it is known that the black kind will flow more abundantly than that. Moreover, the black is of this kind that it cannot be burnt in a lamp, unless the reservoir is thoroughly sealed up, otherwise when it is brought near the flame, the whole of the oil catches fire.

Indeed, on the contrary, on account of this difficulty it is not usually put in lamps. The fact that the price of black oil is low is to be ascribed to the fact that supplies of it are ample, and it smells of asphalt more than the rest. The white kind is kept separately by the tramps for painters. The other two, the red and the black, they sell mixed together, so that you will hardly find it anywhere in a pure state. The greatest deceit is practised by the wandering beggars and pedlars, of which there is a large number throughout the cities of Italy, and who have now as far as into our own land of France, so that now you will see many people burning the oil in Lyons and Paris in the market-place, in the streets, and in the palace. They dip a knife in it, hold it near a fire, so that it catches light, and thus they cause the oil to burn and shed light: because however much is dropped by them, falling from the knife above on to an outstretched hand, nevertheless the hand remains unscathed. This fact which caused the multitude to wonder as if it were in some way a new portent, and diligently enquired what this thing might be. Then the pedlars, being made for tricks and deceit, promised all sorts of things about their oil, and for this purpose they had ready small phials full of naphtha oil: wherefore, where they extolled to the bystanders its virtues, it was much easier for them to induce them to buy.

Since now I have written enough about the kinds of naphtha which are popularly known, it seems to me that something should be added about oil in Bulgaria: indeed, I knew it when I was in Byzantium. The oil is black, fluid and pungent in odour: and this burns in the same way as naphtha: and from this I was at first easily tempted to believe that it was naphtha: but I straightway changed my opinion for a certain reason. For I discovered that the naphtha was for sale and hence unlike oil. But this has the most abominable smell in the whole world—something like asphalt. Whence I suspected that it came from Iuniperus (berries). Some of the Jews maintain that it flows from a mountain, and they ascribed greater power to it than to Naphtha. But when I learnt that it was less greedy of fire, then I ceased to believe that it was naphtha any longer. It is a very great marvel, and to be considered as supernatural, and like to an omen, and worthy of a wonder, that indeed men believe that they can imitate things which are of a single nature, in such a way that they deceive even the most skilled investigator of facts that it is permissible to regard this naphtha as oil, inasmuch as they adulterate it with foul and very evil-smelling water, so that it often misleads the unwary. Which fact I should never have believed to be true, if I had not thus discovered that this was actually the case. However, since indeed nearly the whole of France uses this hardly thus adulterated; for this reason, I should like to relate what is known about each, so that the adulterated kind can easily be distinguished from the genuine kind. When genuine naphtha is held near a fire, it catches light instantly, which the adulterated kind does not do. For if you place a little of this counterfeit oil on a knife and hold it over a fire, you will see the fire recede, nor will it attract the flame, of which genuine oil is by nature very greedy. Because if it were to catch fire by chance, you would find that very black smoke was given off, and that it smelt of turpentine. Medea burned the mistress (of her husband Jason), the daughter of Creon, in naphtha oil, after she had approached the altar in order to sacrifice (as Pliny writes) her crown caught fire. Whence the Greeks call naphtha "Medea's oil"; Strabo writes that for the sake of an experiment naphtha was poured over a boy at the baths, and a lamp was placed nearby, as a result of which the boy was so badly burned, that he would have been killed outright, if the servants had not mastered the blaze by pouring great quantities of water on it, and thus they saved the boy. Ammianus Marcellinus relates that the Persians coated their weapons with this same oil, and having set light to them and hurled them far and wide, set fire to the enemies' homes. The bitumen springs, and indeed famous ones are found in many places all over the world, such as in Germany near Brunswick, the famous town in Saxony (!), and near the monastery in Swabia (Bavaria!), whose name is Tegern: in a valley in the Jura, whose name Gersefori (Görsdorff, Alsace) came from the word for a hare. In greater Asia, too, there is an abundance of hot springs close to Tralles near the river Charocometes, close to the town of Nissa: which are all so oily, that those who bathe in them have no lack of oil. Similarly there are hot oil-springs close to the village of Dascylium (near Bursa, Turkey). There is besides a bitumen spring in Cilicia close to Soli (near Tarsus), which to this day turns the river Liparis oily, that the bodies of those who swim or bathe in it seem to be coated with oil and from this the name is derived. In Bactria above the region near the river Oxus (Afghanistan), in Media (Ecbatana), in India, in the region of the Troglodytes near Carambis, in Ethiopia, and near Carthage in Africa. Hard bitumen (sic!) is called Obsidian, because it was found in Ethiopia by Obsidius, and from it they make the gem-stones which moreover are called Obsidionae. Pliny says that naphtha-oil is gathered with tufts of reeds, because it very readily adheres to them: Georgius Agricola writes that the Germans gather it with the wings of geese, or with small thin linen shreds if there is not much of it, but if there is a lot of it, they draw it out with vases. The power of the fire present in it is so great that any body which is anointed with it, if it is held near a fire, is consumed, nor is it put out with water, unless in exceeding quantity: nevertheless, it is stifled with mud, earth, dust and by anything completely dry. But whereas it easily catches fire, it is an old-established custom there to use it for lighting lamps, instead of oil, and it is customary in many places, such as in Sicily, on the plain of Agrigentum, whence it is called by some "Siculan oil". In Saxony, country people not only use it to this extent nowadays for lighting lamps, but hence they make bridal torches, having dipped in oil the stalks of dried mullein plant, as Agricola says, and with it they grease the axles of their chariots. Formerly the statues in Rome were painted with bitumen, according to Pliny, book 34, chapter 4. For he writes that the ancients usually painted statues with bitumen, wherefore, strange to relate, it pleased them more than to have them covered with gold. It is evident that the stone gagates is of a bituminous nature, therefore I thought that something should be added about it, after discussing pissasphaltos or Jew's pitch. Galen wrote about gagates in these words: "I brought back from Coele-Syria many stones of the tables, broad in size and black, and which when they were laid on the fire, gave off a thin flame, produced from the hill surrounding the sea which they call the Dead Sea and to wit from the source from which the bitumen is also produced. Moreover, the smell of those stones was like that of bitumen. But without doubt the hot springs flowing from deeper down beneath the said bituminous hill wash away the bitumen and melt it, and pour it into the place from which it sprang. Because indeed black bitumen is found on the sea-shore washed up by the tide; it is true that it is Jew's pitch (pissasphaltos) and only differs from it by its hardness. It is not broad and easy to split, and likewise in this it differs from fossil formations." Thus much he says. As far as amber, or lyncurium (or Pterygophoros) is concerned, I decided after all to write nothing at this juncture, but that according to Herodotus and Diodorus for instance it comes from the firs as I shall prove in fact elsewhere with ample evidence. Also it is false that it is obtained from the ground, whence it is said to gush forth. Forsooth it sinks, so that therefore it is washed up by the waves on the seashore. Consequently, if it comes from the gum-tree (or fir), a description of it has nothing to do with this treatise."

Here ends Belon's discussion of the crude oils he knew; the next chapter deals with saltpetre. It is clear from the text given above that Belon had personally visited several of the oil springs of northern Italy, probably in the course of 1544, 1546, 1549 or 1550, and possibly on several of these occasions. His Latin is not quite up to Ciceronian standards, and like so many of his contemporaries he does not always render personal and place names properly; witness the fact that even official documents did not give standard spellings in those days. Hence some of the seepages in the Parma-Modena region defy closer identification and a more thorough knowledge of local topography would be required to locate them. Several such as Monte Zibio and Sassuolo are of course well-known in early petroleum literature.

One of the merits of Belon's description of the Italian crudes is that, contrary to the usual practice, it is not copied from earlier authors such as Ariosto (1460) but truly his own. In fact we can trace in it the origin of many stories in later accounts, such as the example of the inflammability of oil illustrated by the explosion occurring when a man descends into the well with a lighted lantern [1]. The different methods of skimming the oil from the surface of the brine given by him are well-known from other parts of the world. His remarks on the price of the crudes and their relation to the economy of working up the brine into salt are noteworthy. Clearly he was very much interested in the manufacture of salt, as he has left us such a detailed description of the lute used by the salt-makers to repair leaks in their cauldrons and to seal joints in their apparatus.

Belon's description of the red, white and black naphtha shows that he recognised the great variety of crudes which even one single region may produce. If he encounters a petroleum which he does not know and which has a different smell from those with which he is familar, he becomes suspicious. Thus he suspects the "oil from Bulgaria", probably a crude oil from a Rumanian seepage, to be a concoction distilled from juniper berries in order to imitate the other types of petroleum sold by pedlars on the markets of Italy and France and already subject to much adulteration. Strangely enough, he does not mention the crude of Gabian (France), which was certainly marketed in his day.

In his day the oil gathered on the coast of Agrigento (Sicily) seems already to have had some importance. His description of the crudes and bitumens of the Near East shows that he has personally seen many of these (the Dead Sea and seepages in Asia Minor) and that he knows his classics and the passages which deal with petroleum. Though he is of course mistaken in calling obsidian a type of hard bitumen, his remarks on the gagates are quite correct, and he attempts to relegate the many miraculous stories about gagates to the scrapheap by citing a perfectly rational passage from Galen on this type of natural bitumen.

Belon winds up these chapters with a passage on amber, which he held to be the gum of a tree and not a mineral as most of his contemporaries like Libavius or Valerius Cordus (in his Annotations on Dioscorides) claimed it to be. They took resin, amber and camphor to be forms of bitumen [2], or sometimes even claimed that it was an adulterated mixture passed off on Europeans by the Turks and the Arabs [3], forgetting that it was washed up on

---

[1] R. J. Forbes, Studies in Early Petroleum History I (Brill, Leiden, 1958, chapter VII).
[2] Severinus Goebelius, De Succino libri duo (edit. Gesner, 1565).
[3] Pierre Braillier, Déclaration des abus et ignorances des médecins (Paris, 1557, pp. 83-85).

the coasts of the Baltic, possibly from bituminous springs in the sea as some believed [1]. In his Observations [2] Belon again claims that yellow amber is not a mineral but the gum of a tree and that it has nothing to do with bitumen or petroleum.

We must conclude from the passage reproduced above that Belon gave us a survey of bituminous substances striking for its variety of observations and information and containing many details which we do not find in Agricola and other sixteenth-century authors on petroleum. Belon is less credulous than his contemporaries; his wide experience in many countries brought him into contact with reality and allowed him to observe many things personally which others knew only from books. We should be grateful for the survey of petroleum which he left us in his Admirabili Operum Antiquorum, which shows that in many parts of Europe, including Italy, seepages were exploited in the sixteenth century and primitive applications of petroleum had already taken sufficient root to make it a commercial commodity at many fairs and markets.

---

[1] Juan Fragoso, Aromatum fructuum et simplicium (1572).
[2] Pierre Belon, Observations de plusieurs singularités et choses mémorables trouvées en Grèce, Asie, Judée, Égypte, Arabie et autres pays étrangers, rédigées en trois livres (B. Prévost, Paris, II, 72, fols. 134 v.-135 r.).

CHAPTER TWO

MATTHIOLUS AND PETROLEUM

Having studied Belon's views on petroleum and petroleum products it will be interesting to compare them with those held by Matthiolus, a contemporary of Belon and one of the most influential pharmacists of his times. Pierandrea Mattioli was born at Siena on March 23rd 1500 and after studying medicine he became honorary physician to the emperor Ferdinand and later to Maximilian II. He was a practising physician at Rome, Siena, Trente and Görz but after having achieved a high degree of practical skill he gave up medicine and devoted himself solely to natural science, more particularly to botany in which he became an expert known all over Europe. He was particularly known for his work on the pharmaceutical application of plants and herbs which he incorporated in his editions of the work of Dioscorides, the Roman pharmacist. The two main works on Dioscorides [1] went through many editions and were published in many languages. Matthiolus was in correspondence with experts and herbalists all over Europe and he took great care to incorporate the latest data in every edition of his works, more than 32,000 copies of which were sold during the sixteenth and seventeenth centuries, a fantastic figure for those days. In 1577 he died of the plague at Trente.

Matthiolus had not travelled as far as Belon, he had seen the oil seepages of Modena and Northern Italy, as well as those of Seefeld and Tegernsee, but further information on this matter was given from hearsay only. Still, being a recognised authority on pharmaceutical matters, his opinion carried great weight, even if he clung to his authority to attack Belon's practical experience and sound judgment.

We will now reproduce the passages from Dioscorides dealing with members

---

[1] Pedacii Dioscoridis de materia medica libri sex interprete Petro Andrea Matthiolo cum ejusdem commentariis (Valgrisius, Venice 1554, 1558, 1559, 1560, 1563, 1565).

Di Pedacio Dioscoride Anazarbo libri cinque della historia et materia medicinale tradotto in volgare da M. P. A. Matthiolo, ecc. (Brescia 1544, Florence, 1547, Venice 1548, 1552; Mantua, 1549).

of the bitumen family [1] and add to each of these passages the comments given by Matthiolus [2]:

"Pix liquida: French, Poix liquide or fonduë: Greek, Pissa hygra: Arabic, Eerf, Cest, Zest or Kir: Italian, Pece liquida: Spanish, Pexnegra: German, for all types of pitch, Bech."

"I. 94. Pix Liquida, which some call Conum, is gathered of the fattest wood of the Pitch tree and the Pine tree. That is reckoned the best which is glittering, smooth & cleane. It is good for lethatlia medicamenta, for the Phthisicos, for ye purulenta excreantes, for the Tusses, for the Asthmata, & for the humora in the Thorax which are hardly excreable, being taken with Hony in a Lohoc to the quantitie of a Cyathus. It is good also being anointed on for the inflammations of the Tonsile, & the Uva and ye Anginae, as also for the aures purulentae with Rosaceum, and for the biting of Serpents being layd on with salt, ground small. Being mixed with the like quantities of wax it drawes off rugged nayles, and it doth dissolue the tubercula vulvae & the Duritias Sedis.

Being sod with Barly meale & the urine of a boy it breakes the Strumas round about: with Brimstone, the barke of the Pine, or Bran, being anoited on, it stops Serpentia ulcera. Being mixed with the Manna (of Thus) & Ceratum it conglutinates Sinuosa Ulcera, & is good for the Fissurae pedum et Sedis, being anointed on, & with Hony it fills up ulcers, & doth cleanse them. With raisins of the Sun and Hony, it doth Circumscarificate Carbuncles and rotten ulcers. It is also profitably mixed with Septick Emplasters."

"Picinum oleum: French, Huile de poix: Arabic, Kepsen or Kapse: Spanish, Azei de Pez: Italian, Olio d'ella Pece.

I. 95. Of Pitch is made, oleum Picinum, the watry matter of it which swims on ye top as whey doth of milke, being separated. This is taken away in ye seething of the Pitch, by laying cleane wool over it, which when it is made moyst by the steam thereof ascending up, is squeesed out into a vessell and this is donne as long as the Pitch is seething. It is available for the same purposes as liquid pitch. Being layd on as a cataplasme with barly meale it doeth restore the hair fallen by ye Alopecia. The liquid pitch doth also cure ye same, & being anointed on it doth also cure the boyles and scabs which are upon cattell."

"Fuligo liquide picis: French, suye de poix liquide: Italian, Fuligine della pece.

I. 96. The Fuligo of moyst Pitch is made thus. Having lit a new lamp put a match of some pitch into it, & covering the lamp with a new earthen vessell made in the manner of a Clibinus, and below having a mouth as ovens have, & soe let the lamp burne; and when the first liqour is spent, put in other, till you shall have fuliginated Fuligo enough, & then use it. It hath a sharp & binding facultie. It is to be used amongst medicines to make the eye-lids faire, and for Circumlitiones, and when haire is to be restored to eye-lids overcharged with watery humours; and it is good for weake and weeping & exulcerated eyes."

[1] We quote from: "The Greek Herbal of Dioscorides", Englished by John Goodyer (1655), edited and printed by R. T. Gunther (Oxf. Univ. Press, 1934).

[2] Matthiolus' notes have been translated from the French text prepared by Antoine du Pinet and published by Jean-Baptiste de Ville at Lyons in 1680.

"Pix arida, Dry Pitch, Poix sèche.

I. 97. The drie Pitch is made of ye Liquid being decocted. It is called of some Palimpissa (as we should say, Pitch boyled againe). There is some of this that is clammy like birdlime, called Boscas, & an other sort that is drie. That is good which is pure, and fatt, and well-smelling, and subrutila and Resinosa: such is the Lycian & the Brutian (from Pinus brutia Tenore) partaking of the twoe natures of the pitch and also of the Rosin. It hath a warming mollyfying facultie, pus movens, discussing of Tubercula and Pani, & ulcera replens. It is profitably mixed with vulnerarie medicines."

"Zopissa.

I. 98. Somme call the Rosin (which together with the wax is scraped off from ships) Zopissa, which by somme againe is called Apochyma, it being of a dissolving nature, because it is macerated in sea water. Somme have called the Rosin of the Pine Tree by the same name.

Note: Although Dioscorides talked specifically in various chapters of several kinds of pitch and their oil and soot, seeing that these are all so well known there is no need to repeat them and I have thought it better to leave them out. However since some people may perhaps take pleasure in hearing how pitch is made I shall set down here to satisfy them and also myself what I have seen of this process near Trent in the mountains of Fleme. Pitch then, which is known as ship's pitch because it is used for coating ships, is made in this way: old pines are selected which are entirely converted into torches and they are split into pieces as if for making charcoal.

Then a floor is made, a little raised and arched in the middle and sloping equally towards the edges. It is paved and cemented with plaster so that the liquid which the pine torch will exude can more easily run into the channel surrounding the said floor. Then the pieces of torch are stacked neatly on the floor in the shape of a pile of faggots used for making charcoal. After this the pile is covered and surrounded with pine branches and equisetum. This done it is covered and stopped up with earth so that neither flame nor smoke can get out. Then it is set alight through a hole at the top, neither more nor less than is done to charcoal. And the flame, having no place for exit, applies the most powerful heat to the amassed wood and on this account the pitch melts and, running down the floor, drops into the channel surrounding it and from the said channel to other channels, well arranged so that the pitch collects in certain hollows in the ground, well lined with boards so that the earth does not drink up the pitch. From there the pitch is drawn and carried away in casks and barrels. The work is known to be over when the pile grows small and no more pitch runs. We have often seen the men in the hills make pitch in this way and they appear to have learnt the way of doing it from Theophrastus, for he said that the Macedonians half burn it in the same way. Galenus, speaking of the properties of pitch, said "pitch dries, warms and dries to a second degree although it is more desiccative than warm and by virtue of the subtlety of its matter it serves excellently those who have shortness of breath and who suffer from the phlegm and who have chilled and impure blood. For this it is enough to take it as an electuary, with honey, to an amount of $2\frac{1}{2}$ ounces. Besides this the above pitches have maturative, astrictive and digestive virtues and are sharp and bitter to the taste. Mixed with wax they cause leprous nails to drop off, clear skin troubles and recurrent rashes and maturate the hardnesses and crudities of apostemes when applied as a plaster in the aforesaid manner. The liquid is more remedial and effective for what has been said above but although the pitch dries less remedially it is more suitable for melting into sores than the other". By this we can see that the liquid pitch is evidently warm and moist. Galenus also mentions the soot of pitch saying this of all kinds of soot: "All soots are desiccative and are therefore of terrestrial substance although they retain a very little of the character of the fire which burned them. That is why all are of a terrestrial nature and subtle parts. The only difference between them arises from the diversity of the materials from which they come, for a burning, sharp material will produce soot of the same quality. Similarly, gentler and more moderate materials produce more pleasant soot. Soot is used firstly as incense for medicaments prepared for eye troubles, even in those made for phlegmons and burning apostemes of the eyes and for their defluxions and ulcers, for it cleanses them with flesh. They are also used in liniments to beautify the eyebrows and reclothe eyelids which have no lashes. The soot of turpentine and of myrrh is no more violent than that of incense. But that of storax seems to be sharper and more beneficial, and still more so that of liquid pitch and most of all that of cedar. The sharpest are used for weaknesses of the eyelids and inflammations of the corners of the eyes and their defluxions, where however there is no phlegmon nor burning aposteme. But the more moderate soots are used for all the aforesaid maladies and for all those for which the soot of incense is known to be effective.

In the work of Oribasius, written by hand, we have ἰξώδης which means sticky and viscous and not ἐνώδης which means odoriferous. To know which is the better description a test can be made with dry pitch."

Matthiolus has little news to report on wood-tar and pitch except for his description of the local method of preparing them, which, however, differs little from what is reported from other European countries and which is practically equal to Pliny's story of the production of Bruttian tar. His remarks on the various types of petroleum and natural asphalt are more illuminating for they reflect his knowledge of the natural deposits of such substances. They are appended to the following paragraphs from Dioscorides:

"Asphaltus: Latin, Bitumen: French, Bitume: Arabic, Hafral Ieudi or Chefer Aliheud: German, Iudenleim: Italian, Bitume.

I. 99. The Judaicum Bitumen is better than the others; that is reckoned the best, which doth shine like purple, being of a strong scent & weightie, but the black and fowle is naught for it is adulterated with Pitch mixed with it. It growes in Phenicie, also, and in Sidon, & in Babylon, & in Zacynthum. It is found also moyst swimming upon wells in the countrie of the Agrigentines of Sicilie, which they use for lamps instead of oyle, and which they call falsely Sicilian oyle, for it is a kinde of moyst Bitumen."

"Pissasphaltum: French, Mumie: Arabic, Mumie, Mumiay or Mumia: Spanish, Cera di Minera: German, Tirschenblut.

I. 100. But there is somme called Pissasphaltus, which growes in Apollonia neere to Epidamnus, which is carryed downe from the Ceraunian mountaines by the violence of the River, & is cast upon ye shore, growing into knobs, which smell of pitch mingled with Bitumen."

"Naphtha, kind of Bitumen.

I. 101. There is somme called also Naphtha, which is ye straining of the Babylonian Asphaltus, white in colour; there is also somme found which is black. It hath an attractive power of fire, for that it doth draw it unto itself from a distance; it is good for the suffusiones oculorum, & the Albugines.

Properties of Asphaltos. All Bitumen hath a power or repressing inflammation, of conglutanating, of discussing, of mollifying, effectual for the Vulvae strangulationes, & for procidentias, being applyed, or smelt to, or suffumigated. Furthermore it discovers such as are troubled with the Epilepsie being suffumigated, as the Gagate stone also doth. Being dranck with wine & Castorium, it drives out the Menses. It helps inveterate coughes, & such as are troubled with the Asthma, & difficulties of breathing, as also the biting of Serpents, & the paines of the hipps, and of the syde. It is given also to such as are troubled with the chollick as a Catapotium, & it doth dissolve clots of blood being dranck with Acetum. It is given also by the way of Glister to such as are troubled with a Dysentrie being melted with Ptissana. It cures distillations Being suffited, & being wrapt about them it asswageth the paines of the teeth. But that which is drie, being warmed with a splatter (& soe layd on) doth agglutinate the haire. It helps those also who are troubled with ye Podagra & with the Arthritis, & with ye Lethargie, a plaister thereof being layd on, mixt with barly meale & wax and Nitre. But Pissasphaltos can doe as much as Pix and Bitumen mixed together.

Note: The real bitumen of Judea is not brought into Italy now as far as I know for that used by the apothecaries is an imitation mixture of pitch, petroleum oil or rock oil and other mixtures. That is why we should not be surprised if it does not answer to the description given by Dioscorides. Brocardus, who has described Palestine, says that good bitumen grows in Judea in a certain lake into which the river Jordan flows about 15 miles away from the town of Jericho. But the bitumen is nothing more than a certain fat which is floating on the surface of the said lake and which, carried by the wind and the waves, thickens and is very sticky. In this lake according to Galenus neither animal nor plant grows and nothing is to be seen because the water is extremely salty. This moreover in spite of the fact that two big rivers flow into it which have an abundance of fish and especially that which is near Jericho called the Jordan: nevertheless the fish do not enter the said lake and do not pass beyond the mouths of the said rivers. Furthermore nothing that is thrown into it goes to the bottom but floats on the water like a boat. This is easy to prove by experience for all boats and ships are more easily kept afloat by sea water than by fresh water. For this reason Galenus also speaks of the above-mentioned place thus: "The water of the Lake of Syria in Palestine (which some call a dead sea and others a bituminous lake) is not only salty but also bitter to the taste. The salt which gathers there is also bitter in itself. At first sight this water is whiter and thicker than seawater and is like pickling brine. It will not dissolve any salt dropped into it as it already contains too much salt itself. When someone dives or swims in the said water he will find himself covered as with a fine salt when he comes out and this water is as heavy as that of the sea for sea water is heavier than fresh water. And even if one wanted to dive to the bottom of the said lake one could not do it for the water resists and bears things, not in order to make itself light as a sophist of the ancients tried to say but (as Aristotle said) this comes of its weight by which it raises anything light as does mud. That is why if a man tied hand and foot were thrown into this lake he would not sink to the bottom. For just as ships floating on the sea can carry heavier cargoes without sinking than they could on fresh water those sailing on the dead sea carry much heavier cargoes than they would on another sea. For the water of this dead sea is as much above the weight of other seas as sea water is heavier than that of ponds or rivers, especially as it is completely full of salt which is a terrestrial substance and heavy. Everyone can easily test this by dissolving salt in sweet water for then it will be seen how much heavier salt water is than sweet water. Even when it is desired to find out whether a pickling brine is ready to salt sufficiently a whole egg should be put in it: If it floats on the top the brine is salty enough but if it goes to the bottom the water is still too sweet. But it is too salty if it returns the salt as it is put into it: this has not been able to dissolve because of the large quantity of salt already in the brine. If you weighed this water you would find it the heaviest of all. And for this reason I myself made quite ineffectual the ambition of a rich man from our Italy who had had so much water brought from the dead sea that he had filled a cistern with it; for I put plenty of salt into sweet water and by this means made it like that of the dead sea". That is what Galenus said about it. This lake which some call a dead sea is the one which the holy scriptures say took the place of Sodom, Gomorrah and several other towns which were engulfed and consumed by the fire of heaven. This is also to be found in Galenus who in the place cited says that this lake is called the lake of Sodom. This lake (as a patriarch of Jerusalem has written) gives off continually a stinking fog which, driven by the wind through what were once very fertile valleys, has made them absolutely barren so that since the existence of this lake there is a great amount of countryside where no grass, nor tree nor wheat will grow nor any kind of greenery except around Jericho where the gardens and orchards are watered by the Elysean fountain. Pliny says that this lake is a hundred miles long and twenty-five wide. He also includes mummy under the types of pitch saying "Pissasphalt is formed naturally in Apollonia from pitch mixed with bitumen. Some do the mixing themselves to make pissasphalt. The pissasphalt is still being produced and is brought to Venice in great quantities from Valona, a town in Apollonia, to tar the ships for which it is very suitable when mixed with pitch. It has also been brought lately from Slavonia and is drawn in that country near Fellin and quite near to Narente. I have been able to get some through friends. It has also been found in Hungary a short time ago and the people of that country call it mineral wax. Fuchsius says pissasphalt is found three miles from Innsbruck and is called Tirschenblut in German. He says he still has some pieces obtained through George Collimitius and when it is lit it produces a smell of bitumen and pitch. But I am afraid that Fuchsius and his George are mistaken, for the latter took instead of pissasphalt the stone gagate which is found in great quantity (according to what Fuchsius says) on the banks of a certain stream which is at three kilometres distance from Innsbruck, because although the stone called Tirschenblut in German burns like the stone Gagate, and when burning gives off a smell of bitumen, nevertheless it does not soften in the fire like pitch and asphalt but on the contrary burns like wood or a pine torch. Moreover Pliny says that the last type of bitumen which is called Naphtha is found in Astagene, Media. It is so attractive to fire that the fire will jump across to it from wherever it is in the proximity. And although naphtha is not produced in Italy as far as I know nevertheless a type of it grows in several places in Italy which has the same effect with regard to fire as Median naphtha, as is clearly seen in the petroleum which issues forth near Modena which is called rock oil. Meanwhile I am sure you will find very surprising what was told to me regarding petroleum by Count Hercules at Contrariis Ferrarois when Maximilian was crowned king of the Romans and of Bohemia. He told me there was a well on one of his farms in which, through secret channels and conduits, petroleum extended downwards in the same way as the water coming to the well. He, desiring to prevent such a harmful thing to his well, said that when he saw it he called in a plasterer to have him block up the said gaps and conduits as best he could. The good man came to the scene of the trouble and, asking for a lighted lantern, went down into the well. And he assured me that the petroleum, immediately attracting the fire, set the whole well alight and with a very vehement vapour (as from artillery) it threw the plasterer out quite dead and the cover of the said well into the air, leaving burning bottles full of petroleum around the mouth which did great injury to the onlookers. This makes me think that petroleum is nothing other than naphtha, which Dioscorides and Pliny say is filtered bitumen. But to return to our first line of discussion we have no Bitumen of Judea which is not artificial or adulterated. Because of this Brassavolus, believing the mummy of the Arabs to be the real asphalt of Palestine, said that in the absence of true bitumen mummy can be used. He was of the opinion that the anatomy and these dried bodies, as much Arab as of other nations, which are brought from Syria to Venice and sold as mummy are the bodies of the poor of those nations who, never having had in their lives the money to be embalmed in the manner of the Jews with aloes, myrrh, safran and balm, as the bodies of those who could afford it were treated, were simply embalmed with asphalt, otherwise bitumen. And I believe that Brassavolus has based his opinion on Strabo who said that bitumen from the lake of Sodom preserves dead bodies from putrefaction and corruption.

On Mummy called mumia

However, as far as I can gather from Arabic writings, mummy is pissasphalt rather than asphalt. For Avicenna said that mummy has the same property as pitch mixed with asphalt, by which it is easily seen that the dried bodies of which we have spoken have only been embalmed with pissasphalt. Serapeum is also of the same opinion and, according to Dioscorides, when he speaks of mummy he attributes to it word for word the same qualities and properties as Dioscorides had attributed to pissasphalt, speaking of it as follows: "Mummy comes to the territory of Apollonia and descends from the mountains with the torrents of water and is found on the banks of the said torrents congealed and heaped up like wax. It gives off quite an unpleasant smell approximating nevertheless to the smell of pitch mixed with asphalt. It has a property similar to that of asphalt mixed with pitch. And although Strabo says that in Judea bitumen was used for embalming and preserving bodies from corruption he does not deny however that pitch was mixed with the bitumen in order to make artificial pissasphalt by this means. But Serapeum and Avicenna were too skilled in these compositions because the Arabs embalmed dead bodies with mummy exactly as the Jews and Syrians were accustomed to do. Why am I not able to agree with Brassavolus when he says that ordinary mummy can be used in place of bitumen in all compositions where bitumen is required? Because, apart from the fact that mummy is the true pissasphalt (in my opinion) or asphalt mixed with pitch, it is always affected by the humour of the dead bodies which usually distils in sepulchres, by which as you may well imagine its nature is radically altered and changed. It would be much better to follow Galenus and Aegineta who substitute liquid pitch for bitumen. Furthermore it must be noted that Serapeum added the pissasphalt of Dioscorides to his mummy so as not to separate two things which were so closely linked. For he knew quite well that corpses were filled with pissasphalt to embalm them just as others were embalmed and filled with balm, myrrh, safran and aloes from which real mummy was formed after an interval of time. The said Serapeum speaks of it in this way: "The mummy of the sepulchres is formed of myrrh, aloe and other drugs and also of the humour produced by dead bodies". But today such mummy is not brought into Italy any more because in those regions where it forms the only bodies embalmed with this mixture are those of great men and those who can allow themselves their own private sepulchres, which are well and carefully closed. This makes it difficult to rob the bodies of the rich thus embalmed as they rob the bodies of the poor who are embalmed and filled with pissasphalt or asphalt mixed with pitch. That is why those people are greatly mistaken who take mummy to be the flesh of these bodies dried in this way and not the substance with which they are filled, as do the shop-keepers who crush the bones and flesh of the said dried corpses and put them into the compositions where mummy is required. Thus to have mummy which is guaranteed one would have to fill and embalm the bodies of those who die in hospitals with aloes, myrrh and safran and disinter them when were ready. Pierre Bellonius of Mans strongly contradicts this point of view and thinking himself very wise for having seen Asia, Greece, Syria, Egypt and Arabia he says and believes that there is no mummy amongst Greeks or Arabs other than pissasphalt. His arguments have seemed so frivolous to me that they have not been able to alter my opinion but have rather led me to believe that it is he who is mistaken. I hope to show this more fully in my Epistles where I shall present a host of errors of the said Bellonius with which he has filled the books he has written touching medicaments which preserve bodies from putrefaction and resin-bearing trees and equally his treatise on fish. For in these commentaries we do not wish to linger too long to uphold and defend our opinion in places where we would have been criticised for other ones. I have thought in this work only to restore Medicine to its purety and its first simplicity and sincerity and to cleanse it of all ordure and lies. To return then to our mumia, according to the Arabs it has several great properties for it is warm and dry to the second degree and relieves headaches caused by the cold without any excess matter appearing. It is good for the migraine, for paralytics, for those who have a sour mouth or epilepsy and for fits of giddiness and vertigo put into the nose with marjoram water. A grain weight of it dissolved in the oil of white stock or henbane distilled and applied in the ears is very good for those pains caused by the cold. Four grains weight moistened and dissolved in a decoction of savory is very good against pains in the throat. Drunk continually for three days with a decoction of barley, sebestan and jujube it is a singular remedy for coughs. Four grains weight taken with mint water removes and cures faults and passions of the heart and it gets rid of all flatulence of the internal organs and the ventricle when taken with a decoction of cumin, ammi and caraway. Four grains weight taken as a potion with ten grains of Armenian bole and the root of the madderwort and five grains of safran all with laxative senna does good service for those who have fallen from a high place on to their stomachs and for blows received on the stomach, on the spleen or on the liver. Four grains weight taken as a potion with a decoction of ache and cumin cures the hiccoughs. Put into medicaments for purging the nose with abelmosk, castor, camphor and oil of ben and applied in the nose it cures headaches which are inveterate and even those which cannot be cured by other means. Four grains weight of it taken and gargled with honeyed vinegar is good for the quinsy and drunk with a decoction of caraway it is good for weaknesses of the spleen. If it is drunk with a decoction of truffles or sea-eggs and asafoetida it serves as an antidote and drunk with wine it helps against the stings of scorpions or putting it on the sting with fresh butter made from cow's milk. Applied, it stops any external bleeding and taken as a potion it stops all internal bleeding and thus it is exceeding good for those who spit blood. Boiled with goat's milk it is good for exulcerations of the bladder and for those who cannot hold back their urine. Some esteem that the pulverized bones of the humain body taken in a potion do much against infirmities of the body and say that each bone has an appropriate member to cure. I deem this to be true in part since from my experience I have often seen pulverized human cranium serve well against dizziness, colic with gravel and other kidney troubles. Now it is time to return to our bitumen from which mummy has led us too far aside. Galenus mentions bitumen as follows: "Bitumen also ranks amongst those things which grow in seawater and in other water, not too unlike seawater, such as can be seen from Apollonia to Egypt and other places round about where this drug grows like a foam floating on the live waters. It is soft while it floats but afterwards, dried, it becomes harder than wood. The best bitumen grows in the sea which is called dead. But this salt lake is in lower Syria. As for its property, it is warm and dry to the second degree. That is why it is used with good results to seal up fresh cuts and in all things needing to be dried and moderately warmed".

Mathiolus' notes on the Dead Sea are based solely on reports of the ancients and of a few travellers who saw it and the Dead Sea Asphalt floating in it. It is interesting to note that this natural asphalt (or rather asphaltic bitumen emitted by seepages below sea level) was no longer sold in Italy as it was in classical times. Matthiolus reports that it was held to be a mixture of pitch and crude oil and hence it could easily be imitated locally, he himself, however, doubts whether this is true and points out that such a mixture would differ from the substance described by Dioscorides.

Pissasphaltos he holds with Pliny to be a natural mixture of pitch and natural asphalt. Nature produces it in Apollonia (Albania) and it is brought from Valona to Venice. Matthiolus refers to another type of pissasphaltos from Slavonia and mentions that it is found near Fellin. We know that rich oilshales are frequent on the southern border of Esthonia and maybe local pockets of natural asphalt provided the samples of "pissasphaltos" which friends sent to Matthiolus. He also groups under the same name the substance found in Galicia which is called "mountain wax" (Bergwachs), but he has probably not seen this mineral wax or else he would never have called it a variety of pissasphaltos.

It is interesting to note that Matthiolus was in Innsbruck and saw the "Dirschenblut" of Seefeld, the production of which we discussed elsewhere [1]. Merenda and he were right in correcting Fuchs, who believed it to be a "pissasphaltos", it was the "gagates" of the ancients.

As to the crude oil or petroleum Matthiolus not only mentions the Median (Persian) oil of the ancients but also the petroleum of Modena (Italy) which was achieving fame all over Europe in his days. We note that he too, like Belon, gives the story of the mason who descended into a well with a lighted lamp to repair crevices and who was blown up by the ignited gases, though not in so many interesting details as Belon did. He decides that "naphtha" and "petroleum" are identical.

The greater part of Matthiolus' note is devoted to the discussion of "mummy" [2]. Here he confuses much of his information. It is true that Brasavola had reported in 1536 that the "mummy" brought from Syria to Venice was none but the wellknown Dead Sea Asphalt (Bitumen Iudaicum) and that the merchants often bought bodies of slaves and criminals to fill them with this bitumen and age them in the desert sand to sell the final product as real "mumia aegyptiaca". However, Matthiolus seems to hold that the Arabs, imitating the ancient Jews and Syrians, mummified the bodies of their dead, but this practice was unknown to these three nations, except for very few cases where important persons were embalmed in a manner entirely different from the practices of the ancient Egyptians. He also holds that "mumia" was a true "pissasphaltos", that is a mixture of pitch and natural asphalt, whereas the original "mummy" (or "mumia primaria") was a waxy natural asphalt found in Persia. In view of the confusion already reigning in older handbooks we need not wonder that Matthiolus, without our modern knowledge of the members of the bitumen family and without analytical tools, could not make head or tail out of these reports. However, this does not provide him with an excuse for the vicious attack on Belon which he makes here, because Belon did not believe in the miraculous properties of this "mummy" and held that this useless and even dangerous drug should no longer be used by the pharmacist or physician. Still we must remember that this battle over the properties of mumia raged for another 250 years.

Matthiolus winds up his remarks with the opinion that "bitumen belongs to the substances which grow in sea water", *i.e.* that it is of marine origine. He states correctly that the bitumen fresh from the seepage is more ductile

---

[1] R. J. Forbes, Studies in Early Petroleum History I (Leiden, 1958, pp. 65 ff.).
[2] R. J. Forbes, Studies in Early Petroleum History I (Leiden, 1958, pp. 162 ff.).

(contains lower boiling fractions) and gradually hardens. He then goes on to discuss "gagates" and other special forms of natural asphalts [1].

"Lapis Gagates.

V. 146. But of ye Gagates that is to be preferred which is kindled quickly & savours like Bitumen in ye smell. But it is commonly black & ill-favoured, and also crusty & exceedingly light. And it hath a softening, and dissolving faculty. And being suffumigated it is a discoverer of ye epilepticall, and it fetcheth again women that have hysteria, & it is mixed with podagricall and acopicall, medicines, and it drives away serpents with ye smoke. It grows in Cilicia at a certain fall of ye river flowing into ye sea, & it is near ye city called Plagiopolis. Ye place and the river at ye mouth of which these stones are found is called Gagas (& they say also that it is a discerner of ye holy grief, for when it is smelt being born down to ye earth, they fall on a soudain, & that being perfumed it doth heal ye hidden griefs of women, they taking in ye vapour).

Note: As for the gagate stone, Mr. Jean Pierre Merenda of Brescia showed me one which had been found in a stream near Innsbruck in the province of Tyrol and which conformed in every way to the description given by Dioscorides. For being brought near to the fire it ignited straightway and smelled of bitumen and was dirty, black, crusty and very light. Plenty of it is found in Flanders and Brabant for the local inhabitants having no other wood, warm themselves with this stone. Not long ago some mines were found near Brescia in Italy, for Mr. Sant. Santin, an apothecary of Trent who is very diligent in his art, and lives under the sign of a coral stone, sent me a large piece. Fuchsius (as we mentioned ourselves in the first book) takes the gagate found in the province of Tyrol to be mummy or pissasphalt. But I am of the opposite opinion as I have amply stated in the above-mentioned place in the chapter on pissasphalt. In another place the same Fuchsius takes black amber to be gagate. But the good man is sadly mistaken, since apart from the fact that amber is not stone nor a type of stone, it is neither dirty nor scaly, but clear and smooth on the top. Moreover, several persons think that the charcoal stone which, according to Agricola, are found in many parts of Germany, are real gagate because they burn in the fire neither more nor less than wood charcoal. But considering that charcoal stone does not catch fire without being blown and that it has no bituminous odour, in my view their opinion is inadmissible, for the gagate stone is so full of bitumen that when lit it burns like pitch and gives off a very black smoke. Likewise, when it is burned with fire in suitable glass instruments, yields oil copiously which, according to Mesué is good for demented persons, and for those subject to epilepsy and for paralytics and spastics, and even for those who, having fainted, their whole bodies remain stiff, and for the gouty and even for women labouring under the affliction and those who cannot contain themselves. This oil cannot be obtained from the said charcoal stone to be dispensed for every humour. Galenus, speaking of the gagate stone, says the following: "There is another black stone which, when it is brought near the fire, gives off a smell like that of bitumen. Dioscorides and several others say that it is found on the river Gagates, which is in Lycia, and that this stone takes its name from there. As for myself, although I have circumnavigated all the

---

[1] R. J. Forbes, Bitumen and Petroleum in Antiquity (Studies in Ancient Technology, vol. I, Leiden, 1955).

coasts of Lycia with a brigantine to see all exquisite things there, I have never seen a river called the Gagates. It is true, however, that from Lower Syria I have brought several black stones which burnt when brought near the fire. I found them on a certain hill near the Dead Sea on the Levant side where bitumen is also found, and the said stones were similar in colour to bitumen. And, in fact, I used these stones for persistent swellings of the knees difficult to cure, mixing the stones with other medicaments suitable for reducing such an accumulation of humours, and I found that they were very suitable and fortified the other medicaments. I also added to the medicament a substance called barbarea and in fact I found that this made the medicament more desiccative, so that one could fill and cicatrise cavernous ulcers and fresh sores even more readily, the medicament being very suitable for this purpose". That is what Galenus says of the matter. As for Pliny, he was greatly in error in supposing that gagate stone would catch fire in water and be extinguished by oil, just like Thracian stone, because these two stones had moreover the same nature. This is wrong, for Dioscorides, having said that Thracian stone has the same properties as gagate, then goes on to say that Thracian stone is peculiar in that it catches fire in water and is extinguished in oil. This is not the case with the gagate stone. Moreover, Aetius says that if a person with a weak heart is given a drink of wine in which burning gagate stone has been quenched, that person will suddenly be relieved and cured and will lose the cold sweat which occurs in such cases and will have a normal pulse again. For the colic too this stone is ground up very fine and a dram weight of it must be taken with wine every morning for seven days, and the patient's glass scented during the said time with the perfume of the said stone and one may be assured that at the end of the stated time the patient will have returned to normal health. Furthermore, as the gagate stone reminds me of agate and agates are used for various bodily ailments, I have seen fit to make mention of it here. Agates were first found in Sicily near the river Achates from which they derive their Latin name. This stone is of various colours and has several interlacing veins so that sometimes, having been touched by no brush other than that of nature, several figures of various kinds are found on them. Pliny bears sufficient witness to this when he says "After the sardonyx of Polycrates of Samos great store was set by this stone owned by Pyrrhus who made war on the Romans, for it was said that he had an agate on which were the nine muses and Apollo with his lyre painted naturally and solely by the artifice of nature, and that the colours were so well arranged that each muse with her instrument could be distinguished". And in fact I think it is for this reason that the ancients have given several names to agates, since they are called phassachates, cerachates, dendrachates, leucachates, hoemachates, coraliachates and several other names because there are some which depict pigeons, others horns, others trees, others blood and there are some which are red like coral. Furthermore, Pliny says that agate is good for scorpion stings and regards those of Sicily as being most suitable for this because their vapour prevents scorpions from being venomous in that country. Moreover, he says that the agates of the Indies have the same virtue and that they have many other miraculous properties. He also maintains that merely to look at them does the eyes good and that they quench thirst when kept in the mouth. It is said that those which are reddish like a lion's skin have a greater efficacy against scorpion stings. It is also asserted that in Persia storms are deflected by their perfume and that there they even stop raging waters and impetuous rivers and that their power will be revealed by the fact that when they are immersed in a boiler full of boiling water they straightway cool it. But to render them more efficacious they should be tied together with hair from a lion's mane. The Persians do not value those which are flecked like a hyena's skin or panther since they cause dissension in the house where they are kept. But agates of only one colour make wrestlers and acrobats almost invincible".

"Lapis Thracius.

V. 147. But that which is called Thracius groweth in Scythia (Sintia, in Macedonia), in a river which is called Pontus. It can perform ye same cures that Gagates can, and it is said that it is set on fire with water, and quenched with oil, which is also donne upon Asphaltus.

Note: As for Thracian stone I have never seen it and have never met anyone who has seen it. But Galenus speaks of it in these words: "There is another stone of which Nicander speaks thus: "When Thracian stone is burned in a blazing fire and then thrown into water it will burn, but if oil is thrown upon it, it will be extinguished immediately. The shepherds of Thrace provide us with it and obtain it from a river called Pontus. However, it is not used at all in medicine for Nicander attributed no property to it other than that its perfume keeps makes away". This is what Galenus says about it. I myself cannot believe what is said about this stone and consider it to be a story."

"Lapis Memphites.

V. 158. Lapis Memphitis is found in Egypt by Memphis, having ye bigness of a pebble stone, fat & of divers colours. It is said that this being beaten small, and smeared upon the places that shall be cut, doth bring in then a stupid senselessness, without danger.

Terra Ampelitis or Pharmacitis: French, Terre à vigne: Arabic, Thin alcharin: Italian, Terra Ampelite."

"V. 181. But of ye earth Amphelitis which some call Pharmacitis growing in Seleucia which is by Syria, that which is black is chosen. It is like to small coals of ye Pitch tree, cleaving like a lath, and glistering equally; & moreover not melted slowly when a little oil is poured on it, when it is beaten small. But ye white, & ashy, & unmeltable is reckoned naught. It hath a dissolving and cooling faculty. And it is taken also to make ye eyes-lids fair, & for ye dyeing of ye hair, and for ye anointing of ye vines at the time of their putting out, for it kills ye worms that breed in them.

Note: According to Galenus, ampelite is justified in being so called because the ancients usually made use of it to preserve their vines from caterpillars just as we use bird-lime in Tuscany. It is also called Pharmacite because it is very medicinal, for not only does it prevent parasites from ascending into the vines but it also causes their death. According to Pliny and Posidonius this earth is so full of bitumen that it is thought to be completely similar to bitumen. It is tested in oil for good ampelite dissolves in it. For this reason some think ampelite to be the crusty stone which Galenus said he had found on the shores of the Sea of Lycia in lower Syria, and which gave off a small flame when put in the fire. And these writers base their argument on the fact that Galenus said he had found this stone on a certain hill almost surrounded by the Lake of Sodom and that bitumen fell from this hill into the said lake. And they conclude that Galenus is to be criticised for never realising that this stone was other than ampelite, which, according to Dioscorides, resembles a charcoal of Pesaro. This is why it is not astonishing that it should have the appearance of stone rather than earth. And for this reason too it is to be presumed that there is no great difference between the gagate stone and ampelite earth, seeing that both are produced from earth and bitumen. Moreover, I was recently brought some ampelite earth from Caniola which conformed in every way with the description of it given by Dioscorides."

Like Dioscorides Matthiolus uses the term Ampelites for a host of substances including natural asphalts, jet, natural resins and waxes and bituminous earths and like the "gagates" of the ancients it is very difficult to identify the substances mentioned in Matthiolus' notes. Probably the so-called Ampelites received its name because it could be applied as an insecticide smeared on the stems of the vines and indeed in the Geoponica there are many recipes for similar applications of members of the bitumen family.

Our main impression of these passages is that though Matthiolus could have studied crude oil seepages in Northern Italy himself and though he had seen the Seefeld shales in Tyrol he had not, like Belon, studied these substances closely. He is still too much impressed by the authority of Dioscorides to dare to doubt him on the more important points and he does not seem to have read some of the alchemical and chemical handbooks of his days in which new substances of this type were discussed. On occasions he elaborates certain of Dioscorides' opinions such as those on Mumia to defend him against the contentions of Belon and others who dared to doubt the master pharmacist.

Bound with Matthiolus' treatise our French edition contains a "short treatise on chemistry for the analysis of vegetable, animal and mineral matters, written by a doctor of medicine". It may be worthwhile to quote from this treatise the paragraphs dealing with the distillation and refining of resin and its (cracked) distillates for the author recommends the same operations for the treatment and refining of "asphalt, jet and all bituminous matters" in order to produce distillates and refined products worthy of being used in the recipes given by Matthiolus. This extract gives a fair picture of the treatment to which bituminous substances were submitted in the sixteenth to eighteenth centuries.

CHAPTER THE LAST

Of Succin or Carabé

"Carabé, which is known as yellow amber or succin, is a very pure and well-matured resin or bitumen which runs from the veins of the earth into the sea on which it floats for some time being pushed to the shore by the waves. The sun dries and coagulates it in the form in which we see it. It is found in different colours—

white, yellow or lemon and black. The white is the most higly valued of all, but as it is very rare yellow is with equal success, but black is not used at all. The usual method of preparation is to crush it on porphylry and this is more beneficial than is supposed and in this manner it is successfully administered for spitting of blood, dysentery and the discharges of haemorrhoids, menstruation and gonorrhoes. The dose is from ten grains to half a dram. It is also used by inhaling the vapour in order to reduce the severity of colds and catarrh.

Tincture of Succin

Take four ounces of good succin crushed on the porphyry and after pouring on to it well-rectified spirit of wine until four finger's-breadths of it are floating on the top the mattrass should be covered by any small vessel, and the joints carefully luted. Then it should be matured in a sand bath over a very slow fire and shaken from time to time until the spirit of wine turns a beautiful yellow as a result of some of the succin dissolving. After this, when the desired amount of the coloured liquid has been poured into a bottle and fresh spirit of wine added to the residue and the mattrass has been covered again with its well-luted vessel it should be replaced on the sand at the same temperature and kept there until the spirit of wine has almost dissolved the rest of the succin and is coloured like the first. After this, when the lute has been removed from your vessels and the desired quantity of tincture poured off and mixed with the first, they should be filtered through filter paper. Having put them in a little glass cucurbit, placed in the same sand bath and covered with its head, a small receiver will be fitted on to its nozzle and all joints well luted. About half the spirit of wine should be drawn off over a very slow fire. When the vessels have cooled the remainder in the cucurbit should be kept in a well-sealed double glass bottle, and it can be called a very good tincture of succin which may be successfully used for maladies of the brain and matrix. The dose is from one scruple to one dram in suitable liquids. It is also used with great success in sores and ulcers, and particularly in tendinous parts, mixed in vulnerary decoctions for injections or simply absorbed from a pledget.

On the Distillation of Succin

Take three pounds of coarsely ground succin and put it in a fairly large retort, leaving it half empty, and place it on a sand furnace covered with its cap. Attach a large receiver to it and carefully lute the joints. Apply a very slow heat at first to bring the phlegm out of it and then gradually increase it, and after this spirit, oil and sal volatile will appear all mixed and confused. Increase and continue the heat until no more exudes. Allow the vessels to cool and remove the lute from the receiver. In the retort you will find a black substance resembling asphalt. Put about two pounds of hot water into the receiver and shake well with all the substances in it so as to dissolve the sal volatile adhering to the walls of the receiver or mixed in the oil. Then pour the whole into a phial and separate the oil from the water containing the spirit and the sal volatile.

Rectification of Succin Oil

Mix and incorporate the oil separated from the other substances with as many sifted and washed cinders as it can absorb and put this mixture into a glass retort and distill it in a sand bath over a fairly slow fire. The first oil which will issue forth will be fairly fine and clear and must be kept separate for internal use. Continue and increase the heat gradually to make another oil rise from it which will be a little more coloured and will be followed by another reddish-brown one. When nothing more comes out of it, stop the fire and keep the oils apart. The first is excellent for apoplexy, epilepsy, paralysis and all diseases of the brain. It is also a wonderful remedy for all hysterical complaints and the retention of urine. The dose is from three to ten drops in some suitable liquid. The other two oils should only be used in ointments and plasters to strengthen the nerves and dispel tumours and also for reanimating paralysed parts.

Sublimation and Purification of Sal Volatile from Succin

The aforesaid liquid separated from the oil and containing the phlegm, spirit and sal volatile of the succin should be filtered to separate it from any oily substance, and put in a long-necked mattrass. Pour into it dropwise a good spirit of salt which will cause a great effervescence because of its action on the sal volatile of the succin. When the effervescence has ceased, put the liquid into a cucurbit and cover it with its alembic. Distil about two thirds of its excess moisture which will be an insipid water, after which the temperature should be raised one degree to sublimate the salt which will rise and stick partly to the cap and partly to the upper part of the cucurbit. Allow the vessels to cool and collect carefully the sal volatile which will be very subtle and penetrating. If it is desired still more subtle it should be mixed with an equal amount of purified salt of tartar and the mixture put in a small cucurbit with its cap on to sublimate it on a sand fire; and the salt thus resublimated will be pure and white like snow and should be kept in a perfectly sealed phial, for it is so penetrating and so volatile that it is difficult to keep it long. This salt is a very good diuretic and an excellent diaphoretic and is therefore given for all obstructions of the body. The dose of the first is from fifteen to thirty grains and of the second from five to twelve or fifteen in some suitable liquid.

This distillation may serve as an example for those of asphalt jet and all bituminous substances of which I will give no further examples, not wishing to extend this summary needlessly."

It will be interesting to compare Belon's and Matthiolus' data with those of the third great botanist and pharmacist of their generation, *Valerius Cordus*. His name we mentioned when discussing Belon's travels for Cordus was born at Simthausen (near Marburg, Hessia) in 1515 but achieved fame at a very early age and he was Belon's teacher at Wittenberg. Together with Belon he travelled through Central Europe and Italy and in fact exerted himself so much in travelling about to collect herbs and plants that he died young at Rome in 1544. If we look for a description of the members of the petroleum family in his books [1] we shall be disappointed. In the 1598 edition of the Dispen-

---

[1] His main work was his Dispensatorium Pharmacorum ...... published at Nüremberg in 1546. We cite from the Lyons edition (1561) and the edition published by B. Kaufmann, Nüremberg, 1598.

His second work Annotationes in Dioscoridem .... (Nüremberg, 1546) was not available to the author.

satorium he uses petroleum for the preparation of Diacastoriu Nicolai and states that "Petroleum hoc loco sumendum est clarum, flavum atque liquidissimae tenuitatis, non nigrum aut crassum" (page 52). Later on (on page 245) he defines "petrelaeon or petroleum" as "the naphtha of the Babylonians, which the Italians call "olio de sasso". The best is the white kind, the yellow is less good, both are found and collected in Italy. The thick, black kind is considered inferior, it is found in the region of Bochdorf and Brunswick in Saxony". It is disappointing to find no further data on the oil of Tegernsee, Seefeld or Gabian, not even the traditional data on the bitumens of the Near East culled from the classical authors. It would seem that Cordus never personally visited an oilfield (or as the Gregorian University of Rome terms it "regio bituminosi olei dives") either in his native Germany or in Italy and he seems to have had no interest in the pissasphaltos, gagates or mumia about which his contemporaries tell us so much.

It is clear from our exposition of the data on petroleum as found in the writings of Belon, Matthiolus and Cordus, that there were two schools of thought. The few who like Belon travelled far and were willing to add from their own experience and observation to the traditional data on natural products and their medical or pharmaceutical properties, were long disbelieved and neglected. The much stronger traditional school to which Matthiolus and Cordus belonged stuck to the body of classical data as transmitted by Dioscorides and, hampered by the lack of analytical tools and methods, were loth to add anything to their stock of pharmaceutical products. Paracelsus and his school who claimed that Nature should be made to yield all its secrets to help ailing mankind were revolutionaries rather than careful researchers and their influence on the knowledge of natural products has been very much overestimated.

This is evident when we turn to the medical-pharmaceutical handbooks of the seventeenth century. We find tradition still strongly anchored in the minds of the leading pharmacists. However, with the growing rationalism during the eighteenth century, experimental and observational data begin to invade the body of classical lore. As an example we will show how this change took place in the medical-pharmaceutical handbook [1] written by Stephen Blanckaert (or Blankaarts) (1650-1714), who taught at the University of Franeker and whose encyclopedia enjoyed a vogue for many generations.

---

[1] The first edition of Blankaarts' Lexicon Medicum Renovatum was published at Leyden in 1690. It was republished at Frankfort in 1705 (to which edition the famous Stahl wrote an introduction), then again at Leiden (1717), Halle/Magdeburg (1718), Leiden (1735) (1756). We quote from the 1735 Leiden (Samuel Luchtmans) edition and that re-edited by C. G. Kühn and published by Schwickert of Leipzig, in 1832.

If we compare the 1735 edition with that prepared by Prof. Karl Gottlieb Kühn of Leipzig in 1832 we can see how new data on the bitumens penetrate into the old traditional body of pharmaceutical data. Blanckaert's encyclopaedia was of course written in Latin, but at the end of each article he adds the names of the product in different languages.

The first entry deals with:

."Asphaltos, Asphaltium or Asphaltum is bitumen, grease or resin of the Macrocosm, as hard as pitch etc. It is seen firstly floating on waters or lakes whence it is driven to the shores and being baked with the heat of the sun, the force of fire or the passage of time it grows hard, and dense and becomes very firm and shiny. It can be softened again with heat, is miscible in the fat of oil and is inflammable. The bitumen of Judea produced by the Dead Sea (therefore called Arphachetes) is preferred to the others. It is superior provided there is a purple shine (*viz*. in the black colour) and that it is heavy and gives off a strong odour, but that which is black and dirty is useless. It may be obtained "by splitting it off" from the Asphalt lake (called the Dead Sea) where much is amassed. That which bears the name Trifolio odorato or bituminoso is so called because it emits a smell of bitumen or asphalt.

Dutch: Jodenlijm. German: Erd-Hartz, Erd-Schwebel, Jüden-Pech. French: Bitume Judaique. English: Jews pitch."

To this information the 1832 edition adds data on the complete or partial solubility of bitumen in ether, absolute and 50% alcohol and caustic soda as well as in essential oils. Sulphuric and nitric acid are said to attack it partly. Kühn is also convinced that the famous mumia, at least the natural kind found in Persia is in reality a natural asphalt and he states so referting the reader to the article on Mumia. He also mentions that natural asphalt is found in several parts of Europe but does not specify which.

The information on Bitumen runs thus:

"Bitumen is said by all to be a substance which when brought near to a flame easily catches fire, is thicker than naphtha or petroleum, very sticky, but fluid enough to flow in its primary state. It liquefies in the heat of fire and does not sink in water or mix with it. It may be mineral or vegetable or artificial, whence the name is general."

Kühhn adds that many products are sold under the name "bitumen" which do not have the proper smell and colour and which have been adulterated with worthless material.

Then follows a reference to:

"Gagate is asphalt made perfect by the force of nature, resulting in it being black, hard, earthy, scissile, polished, strong-smelling and shiny. Owing to its hardness it is called Lapis Gagates or Thracius Nicandri."

Here Kühn observes that the same product is also denoted by names like "Succinum nigrum", "Lapis obsidianus", "Gagas", "Gangites", or Bergwachs, Gagat in German, Jet in English and Jayet, Jais in French. The confusion still rampant in many circles between certain bitumens and minerals is obvious from these remarks.

There is of course the unavoidable reference to:

"Mumia. There are four types of this. There is, of course, the Arabic which is a liquid or a congealed liquid exuding in sepulchres from bodies embalmed in aloe, myrrh and balsam. Another is the Egyptian or liquid from bodies embalmed in pissasphalt. Certainly corpses were embalmed with this and with a commoner kind; so much so that the bodies themselves embalmed in this way are sometimes exposed for sale. The third, Pissasphaltum factitium, is a mixture of pitch and bitumen which is sold as mumia. The fourth is a body dessicated under sand by the heat of the sun. It is made in the region of the Hammonites which lies between the Syrian region and Alexandria where the sands of Syrtis lifted by whirlwinds bury unwary travellers and then their bodies become quite dry and dessicated by the heat of the burning sun. The first kind is the best of all."

The Kühn edition mentions that this fourth type of mumia is called "Mumia alba" or "mumia aegyptiaca" which is a hard, friable, blackish substance with a fragrant odour. Several physicians advise not to use it as they believe it to contain arsenic. All authorities agree that the term is derived from a Persian word meaning "wax". A new entry MUMIA PERSICA NATIVA states that this is a natural waxy type of bitumen found in the foothills and mountains of the Caucasus region and that its virtues seem to be many. It is seldom found in European pharmacies and the author doubts whether it is identical with the Egyptian mumia.

In his days Blanckaert could state only that "naphtha" was a liquid, inflammable white or black type of bitumen, which was sometimes called "petroleum". The 1832 edition definitely says that "naphtha" is the name of the light fractions of petroleum and adds that this word "naphtha" is even used to denote the volatile components of acids or natural products distilled with acids.

The knowledge of sources of crude oil increased considerably during the eighteenth century. This is clear from the entry on "PETROLEUM" or "OLEUM PETRAE" which in the 1735 encyclopedia mentions only the seepages of Modena and Parma, Sicily and Babylonia stating that it is either white and clean or thick and black. The 1832 handbook has this passage:

"Petroleum, or Rock Oil, is a mineral oil or liquid bitumen, very thin, and light, immiscible in water or spirit of wine and floating on the surface of either, readily miscible in essential vegetable oils, more or less strong-smelling, immediately inflammable in the vicinity of fire, exuding at all parts from stones, rocks and earth. It is called by some "ethereal fossil oil". Various kinds are mentioned. The purest is "white petroleum" which is also called "true naphtha" and is found in Asia, principally in Persia but in Europe it has hitherto hardly been encountered anywhere except in the duchy of Modena. "Yellow or Italian petroleum", which is of a considerably thicker consistency, is also found in the duchy of Modena. "Red or Gabian petroleum" is dark red, thicker and more strongly smelling, and is found in various parts of Italy and eastern Gaul. It is more common and most of the local inhabitants use it for fires where it is scattered over the ground. We seldom get the purest. There are people who use this oil against the Pictish colic, osteocopic pains and tapeworms. It removes the irritation of chilblains. Lastly there is another petroleum which is found on certain islands of America. It is called "Barbados petroleum", "Barbados bitumen" or "Indian cedar-pitch oil" and is dark red in colour, like honey in consistency, and is less inflammable, and the local inhabitants attribute exceptional medical powers to it. It is often used in England. German, Steinöl, Bergöl, Erdöl: French, Pétrole, huilede pétrole: English, stone-oil: Dutch, steen, peter-olie."

This gradual extension of the facts about petroleum stands in great contrast with that on PISSASPHALTUS (German: Erdharz; French: Pissasphalte, poix minérale) which is carried over unaltered from the 1735 into the 1832 edition. It runs:

'Pissasphalt is natural or artificial. Dioscorides describes the natural kind which springs forth in Apollonia of Epirus. That which comes down from the mountains of Epirus is carried away once more by the current of the river and is cast on to the bank by the waves, where it hardens. And on the account it appears to be no other than asphalt (for which see the chapter on asphalt). The artificial variety is a mixture of pitch and bitumen (whence it derives its name) and appears to be that which our people call asphalt or bitumen of Judea. There are some who think that this is Arabian mummy, and yet the sound of its name denotes something half-way between pitch and bitumen; it is black earthly and strong smelling; perhaps combined with various fats and dark bitumen it is hardened either artificially or naturally."

Thus more scientific appreciation of the characteristics of the members of the petroleum family was growing among the pharmacists. Still folk medicine and magic retained their grip on this art for many a century and some pharmaceutical handbooks did not rise above the level of the medieval manuscripts such as the earliest document in the English language referring to petroleum. It belongs to the times of King Alfred (who died in 899 A.D.) and it was commented upon by Grattan and Singer [1] in these words:

"It is appropriate to refer here to the passage containing the name of the only identifiable contemporary figure mentioned in the Anglo-Saxon medical texts. One

---

[1] Grattan and Singer, Anglo-Saxon Medicine, pp. 50-51; E. O. Cockayne, Anglo-Saxon Leechdoms, vol. I, pp. 288-290.

of the Leechbooks gives a group of recipes which "Dominus Helias, Patriarch of Jerusalem, ordered to say to King Alfred". In the time of Alfred Jerusalem was in Moslem hands and its Greek bishop, Elias III (in office 879-907), was sending begging letters to Western rulers for financial help to restore the ruined churches of his depressed communion there. Bishop Asser, Alfred's biographer, saw such letters addressed to his master. The recipes of Elias were perhaps in return for favours that he hoped to receive or had received. "For apparitions and delusions", Elias recommends, "smearing with *balsam*; for one whose speech fails, *petra oleum* and a Christ's cross marked under his tongue; for a man out of his wits, *petra oleum* and a Christ's cross on every limb; for inward tenderness, *tyriaca*". Here is the first mention in English of *petroleum* and of *tyriaca*."

Generally speaking therefore we find but slow progress in the identification of bitumens and their properties in pharmaceutical handbooks which were apt to cling to traditional lore going back to the classical authors which had provided them with so much valuable information on animal and vegetable products. The advance was to come from the chemists and geologists.

CHAPTER THREE

THE CHEMISTS AND THE COMPOSITION OF PETROLEUM

For several centuries chemistry tarried behind whilst mathematics, mechanics and physics developed rapidly. Quantitative chemistry cannot be said to begin before the eighteenth century except in such restricted fields as assaying. Hence the earlier speculations on the nature of the bitumen family and its members were mostly qualitative and real insight was only slowly gained.

Systematic thought is given to our problem from the middle of the sixteenth century onwards. Georgius Agricola (Georg Bauer, 1494-1555) was convinced that the subterranean heat was due to the combustion of bitumen (which in his mind also included the members of the coal family) and sulphur. The burning vapours of these substances rising from the depth of the earth formed one of the main forces in mountain building [1].

"For the above causes, therefore, there must be some material to feed and replenish the fire to such an extent and for so long, as it could not last without fuel. But this material is called in question: is it dry and like the earth from the Caucasus which today is called peat or is it oily, like earth full of bitumen, seeing that both are burned? The Caucasians cook food and make a fire on this earth which is cut out of the fens and dried, and according to Pliny, in the frozen North they burn entrails on it. In the same way some are in the habit of burning the lumps made by tanners from what they scrape from hides. And liquid bitumen does in fact burn, as not a few burning places testify. But the material for this fire cannot be dry all the year round and yet it is quickly consumed by fire and extinguished by water so it must need be rich. Indeed, there are divers rich materials produced by the earth, for instance as marl, sulphur and bitumen, so that it might be imagined that the fuel for this fire was derived from them. But marl does not burn, nor any rich earth, unless it is sulphurous or bituminous, and indeed sulphur does burn, but is just as quickly extinguished by water. None of these can be the material of this hidden fire. From this it follows that bitumen must be its fuel, since the latter burns in water and feeds on this moisture, thus lasting as long as possible. This is the cause of the apparition of those wonderful fires which we see joined together in water and not extinguished. They always contain in their composition some bitumen or Jew's pitch or substances

---

[1] G. Agricola, De Ortu et Causis Subterraneorum (Froben, Basel, 1558, Lib. I, pp. 12-13, 32, 40).

made of bitumen and we think that the burning stones in water in the mountains of Ephestos, Lycia which Pliny mentions are bituminous, as are also sands. Nevertheless I do not deny that dry sulphur, if it is the substance it is frequently said to be, burns as well as bitumen at the sides of canals and in those burning places that belch

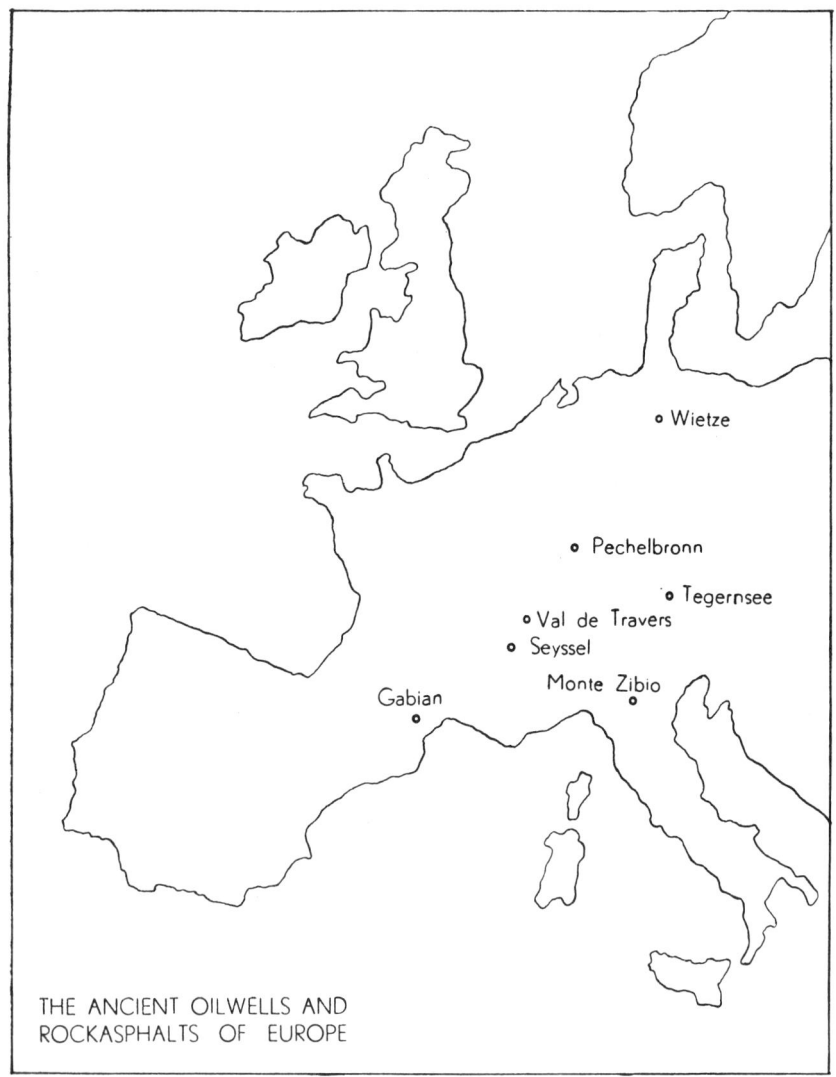

Fig. 1. The ancient oilwells and rockasphalts of Europe.

forth fire, and that in some districts waters are used for producing heat. And indeed hot water is generated in the earth in this way. Actually it becomes lukewarm either because, as soon as it is heated in the interior of the earth, it flows a long way through channels before leaving them, or else such a great abundance of water flows towards the fire that the latter is unable to heat the water to such an extent as to make it boil. But if the warm water flows through channels for a very long time it will not only be cooled but the unpleasant taste, offensive smell and dirty colour which it has acquired will remain with it. But it is rather this same fire that heats the liquid, for a substance will grow hot which was previously pressed out of the earth by the force of heat. But enough about heat and its influence on liquids; I will now turn to its colour. The purer the water, the whiter is its colour, and the more it is mixed with liquids, earths and fragments of stones, the more its colour will differ. This is the reason why hot waters, because they are more mixed, are usually blacker and less clear than others. Thus the water appears black to us when it is mixed with a black fluid, and yellow when mixed with yellow mud of other colours when mixed with others, just as fluids take on the colour of earths or stone fragments with which they are mixed. There are various reasons why the fluids are coloured, for heat first causes them to change from white to red, from red to yellow, from yellow to green or black: in the same way blood is sweetened from phlegm by digesting food, when yellow bile is expelled from the blood by excessive heat and bile is changed from yellow to green or black by combustion. For example, a certain liquid bitumen is white because it is forced out of the earth by a gentle heat; hence white amber is soft because it consists of such bitumen. It is yellow on account of the greater heat; for this reason a certain pungency affects the taste. When black it has most effect of all. For the earth pours it forth just as pinewood under the stress of fire will pour forth black pitch; because the resins which the sun draws out of this kind of tree are for the most part white or yellowish or honey-coloured or red but rarely black. ..... it had expelled bitumen. Lastly, the carbon hill of which I spoke a little while ago is full of bituminous earth. However our people call it carbon because blacksmiths use it instead of carbon on which, when it is burning, they slowly trickle water, as they do on the other types of carbon, so that the fire does not quickly consume it, and it burns more brightly. But sulphur takes second place. For of all materials that are mined none catches fire more easily than bitumen and sulphur because of their oiliness. It is true that burning sulphur is extinguished by water and is quickly consumed, so that burning places belch it forth and it burns in them: the reason however is not the perpetuity of the fires but rather bitumen burning in caverns where there is no lack of water. Just as by adding oil to fire the flame is fed, so by pouring water on burning bitumen the fire will not be extinguished but rather increased."

The friction of internal winds is a minor force as compared with that of the burning bitumen and sulphur according to Agricola. He then devotes the rest of this essay to the origin and distribution of ground waters and juices and to the part which subterraneous "air", "vapour" and "exhalations" play in volcanic eruptions and earthquakes. Another essay in "four books" and 83 folio pages deals with "the substances which flow from the earth",

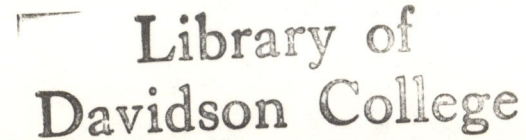

such as water, bitumen and gases. Apart from some data, mostly culled from ancient authors, on waters and wells entraining "liquid bitumen which separates fairly rapidly" and some remarks on the external characteristics of certain crudes known to Agricola and the classics [1] its contents are rather disappointing as the following extracts will show:

"The color of liquid bitumen varies. Some is white, especially the genus naphtha. Posidonius believed this material to be liquid sulphur. It flows from the earth in Babylon near Demetriade, in Parthia, in Mesopotamia and from a mountain they call Gibius on the Mutinian plains. It also flows into the German Sea and from it "white amber" is made. Some is whitish gray as is that found at the port of Sichres, Arabia. The Arabs call this material "solidified amber". "Solidified amber" is formed from a bitumen of a similar color which is dug up on the shore of the bay of the Sudini. Some is reddish yellow such as that flowing from a spring near Tegernsee in Suebia. It is also found at Salachites, India. Reddish bitumen is found in the fields around Modena. Wax-yellow, honey-yellow, and bitumen the color of Falernian wine is dug up from the German Sea. These are colors common to amber. Black bitumen flows from many springs in Germany, in Saxony two miles from the town of Brunon on the road to Scheninga and in the swamps three miles from Burgedorf; on the border of Apollonia near Nymphaeum; and in Babylon. Reddish black bitumen comes from a spring at the foot of Mt. Dester, fifteen miles southwest of Hanover. This bitumen floats on the very clear spring water. That from Judea is a bright purple and some from near Brunon is blue.

Bitumen has a variable taste and in this way we can distinguish if from amber. The black, when entirely dry, has a somewhat bitter taste while the white is oily sweet. Bitumen of other colors is less sweet. It has a variable odor. The black usually has a heavy odor, the white a pleasant one and material of other colors have odors between these two, sometimes almost the odor of myrrh. A black bitumen flowing from a spring at Cratea, Carthage, is reported to have the odor of citron. That from Nymphaeum Apollonia, has an odor of pitch mixed with bitumen and for that reason was given the name of πισσασφαλτας.

Sometimes the material which flows as a spring is cold as at the German localities mentioned above, sometimes it is warm as at Nymphaeum, and sometimes hot as at the asphalt lake in Judea where it flows out at irregular and unpredictable times.

All bitumen is not the same consistency. Some is as fluid as oil as that from Suebia while some is the consistency of mud as that at Samosata which is called maltha and judaicum.

Since bitumen is an unctuous liquid it is light and floats on water like oil."

Agricola tries to describe the colour, taste and other characteristics of these crudes, but he must necessarily remain vague about such seepages as he has not personally inspected:

---

[1] Georg Agricola, De Natura Eorum quae effluunt ex Terra (Froben, Basel, 1558, pp. 89, 93, 94, 97, 100, 101, 104, 108, 109, 110, 118).

"Bitumen floats on water and when it occurs in abundance can be collected in vessels. When it is in meager quantities it is collected with goose feathers, fine meshed linen cloth—a method known to Pliny—and thin mats made from reeds. It adheres readily to these materials. Bitumen contains a very powerful fire essence so that when any material is saturated with naphtha and placed near a fire it will burn strongly. Water will not extinguish it and only seems to make it burn more. It can be smothered with mud, earth, powder and any wholly dry substance. Since it burns so readily it is widely used in lamps for illumination instead of the older olive oil, as for example, in the province of Agrigento, Italy, hence the name "Sicilian oil"; near Solo, Cilicia; and in Babylon, Ecbatan, India and Ethiopia. The country people of Saxony use it today for illumination and for making funeral torches by dipping the dried stalks of mullein in it and also for greasing the axle-tree of carts.

Medea, according to Pliny, burned herself with bitumen when, after making a sacrifice, she drew too close to the altar, fascinated by the burning circle. For this reason the Greeks call naphtha "oil of Medea". This genus of bitumen is quite inflammable and Strabo writes that Alexander experimented with it by having it poured over boys and lighting it so that the boys were set on fire and died at once unless servants poured large quantities of water over them and put out the fire. According to Ammianus Marcellinus the Persians spread it on cloth and after lighting it used it to burn the homes of the enemy as the flame slowly spread over the cloth . . . .

It is spread on copper and iron to prevent rust and corrosion. Saxons paint wooden posts with it to protect them from rain. For the same reasons it is customary to spread it on statues . . . .

It is used in medicine. Drunk as bituminous water it breaks up blood clots and causes abortions. Spread on cattle and beasts of burden it cures mange and Pliny writes that the Babylonians believed it to be good for jaundice and for whitening the eyes. They also believed it to be a cure for leprosy, eruptive and itching skin diseases. It is used as an ointment for the gout. Horses that drink from the Cassinitius River of Thrace are said to become wild and for that reason the water is judged to be bituminous."

Agricola's De Natura Fossilium was his most important mineralogical work, written in 1546 after sixteen years of experience in the mining districts of Joachimstal, Chemnitz and Freiberg. In this essay "in ten books" Agricola tries to classify minerals by colour, weight, lustre, shape, texture, solubility and fusibility which he discusses in the first book.

The second book describes the "earths" such as clay and ochre, books III-V the "solidified juices". We are particularly interested in the contents of book IV which deals with camphor, bitumen, coal, bituminous shales and amber [1]. This fourth chapter opens with the following passage:

---

[1] Georg Agricola, De Natura Fossilium (Froben, Basel, 1558, pp. 169, 170, 222, 223, 224, 225-230, 240); Georg Agricola, De Natura Fossilium (Textbook of mineralogy). Translated from the first Latin edition of 1546 (Geological Society of America, Special Paper No. 63, New York, 1955).

'I shall now take up a second unctuous juice which is naturally related to sulphur and is called ἄσφαλτος by the Greeks. The Latins have named it bitumen. Included under this name are not only the substances the older writers placed here but also naphtha (naphtha), camphor (camphora), maltha (maltha), pittasphalt (pissasphaltus), jet (gagates), Samothracian gem, thracius stone, obsidianus stone and many others classified by Pliny as gems, and natural carbons as well as the earth called ἀμπελῖτις by the Greeks. Amber (succinum) is also included here. This juice is known by so many different names because of variations and qualities by which it is distinguished and because of the discourses of the people in whose countries it either originates or is sold.

First of all the liquid (which people experienced in the nature of things correctly call "liquid bitumen" since it is usually distilled from the solid), being similar to olive oil, is especially unctuous and has been named oleum (oil) by various writers at different times and is now called petroleum (petroleum) because it flows from rocks. This same black juice, when liquid, is called pix (pitch) by others because of the similarity in color to that of pitch. From this it is apparent that the name and nature of this substance was evident and well known to some and obscure and unknown to others. Thus many names have been given to one and the same thing and, at the same time, many more names coming from the vocabularies of different races have also been given to this same juice. The Babylonians called it naphtha, the Samosatians maltha. The Moors, following the Arabs, as did several other races who were influenced by their learning, called it hafral."

Then follows a long passage on camphor, which Agricola believes "is made from a certain genus of bitumen by distillation". He then goes on to discuss bitumen:

"Bitumen which flows from springs is often so dense that it has the appearance of mud. However, as long as it floats on water it remains soft or flexible. When removed and dried it may become harder than pitch. Even though completely fluid after being kept for a long time in a vessel it usually hardens. Dense bitumen is found floating on the Dead Sea and on the stagnant water of the city of Samosata, Comagene. It flows from the Carpathian Mountains at Siebenburg, is found in Rhaetia near Seefeld and in Epirus near Polina. The latter material is called πιττάσφαλτος by the Greeks, a word derived from pitch and bitumen, not because it contains both of them, as Pliny writes (I do not know whose opinion he follows) but because it smells like each of them, as Dioscorides correctly states. The Moors call this material mumia. Serapio gives this name to both this material and the compound used in embalming the bodies of the dead.

There are springs of bitumen on the island of Zante and at many other places. I have described in De Natura eorum quae effluunt ex Terra the occurrences, colors, tastes, odors and other qualities and uses of this mineral and will not repeat myself since nothing would be gained.

Liquid bitumen, having been drawn or collected, is heated in brass or iron boilers to thicken it. When it is finished it usually catches fire but the blaze is extinguished by linen cloths soaked in water and thrown over it. The Germans who live in Dacia and Saxony cook it in this way. I do not doubt that the Deximontanians who, as Pliny writes, live on the right bank of the Granicus river flowing through Susiana treat bitumen in this same fashion. Theopompus has written that the bitumen coming from the crater of Nymphaeum is mixed with some tasteless material and is the most dilute of all. Pliny writes that some pitch is mixed with bitumen and is recommended as a remedy for mange on animals and when young animals have injured the mother's teats. The Saxons increase the viscosity of bitumen by mixing it with old animal fat just as others mix it with pitch. It is dug up in a dense or stiff condition on a hill in Apollonia according to both Theopompus and Posidonius. The former calls it mineral pitch and the latter writes that after removing the pitch the working is filled with earth and after a period of time the earth is changed to bitumen.

All bitumen is unctuous and fire and air are mixed with it in all proportions so that, as a rule, it will catch fire with ease. The more dense varieties catch fire easily when melted and since they possess the quality of denseness they will congeal again when placed in a cool room. Dry bitumen, whether natural or dried artificially, is used in many ways. Semiramis, using bitumen instead of mortar and without doubt before moistening it with water, erected a brick wall around Babylon. The Egyptians used it to preserve the bodies of the dead. The Sabaeans, burning it as incense, inhale the odors as a cure for head ailments. Pliny writes that when burned it is a cure for epilepsy. In medicine it warms and dries in the second degree and therefore coagulates bloody wounds and stops bleeding. Finally bitumen, both within the earth and on the surface, may be hardened and altered until it becomes as hard as a stone.

Let us take up now the earthy mineral to which writers have given various names. Galen called it stone, not realizing that it was the same as the pharmacist's earth he had just discussed. It is called ampelitis earth by those who have written on rural subjects. The Greeks call it Κνιπας because, according to Galen, it harasses the worms gnawing the buds of vines. It is given the same name by physicians and pharmacists because, more than any other earth, it has the efficacious power of healing and curing. Theophrastus calls it carbo because it has the same color as coal; because it catches fire and burns in the same way; and because it is used in the same way as coal. The German name is made up of the words for stone and coal. Actually it is just as easy to make up new words in our language as in Greek.

If bitumen is sufficiently hard to take a polish it is called gagates (jet). According to Dioscorides the name comes from the Gagas river in Lycia which empties into the ocean not far from Plagiopolis. It is found near the mouth of this river. However, Galen, who travelled along the entire coast of Lycia in a small boat and should have seen all the known things that have been found there, writes that he himself did not see the mouth of this river. Other writers have related that there was a town of Gagas in Lycia. According to Stephanus, in his first book on Lycia, Alexander called an ancient wall by that name. It is probable that when the town was deserted the name was transferred from the town to the river. Phocion Grammaticus writes that the Rhodians hid in this town. Pliny mentions Gagas and Rhodiopolis as towns of Lycia, each of which is likely to have been founded by the Rhodians. Nicander the grammarian says that this town in Lycia was called Ganga and Gangis and he refers to jet as ἐγγάγγιδα πέτρος. Strabo calls it gangitis. This same hard polished bitumen is called samothracia (gem of Samothrace) by Pliny—again I do not know his authority—because it is said to be found on the island of Samothrace. Nicander calls it lapis thracius because farmers bring it from a Thracian river that they call

Pontus. An unknown Greek writer states that the Pontus river which carries down this genus of stone is between Scytha and Medos. There is still a question whether lapis thracius may be the same stone as that which Theophrastus calls spinus or the same as another stone which I shall now describe. Each of these stones is bituminous but the latter is simple, spinus complex. I shall discuss the complex minerals in the tenth book.

Obsidianus lapis is a hard bitumen which is so named because it is found in Ethiopia near Obsidius. The gem called obsidiana is made from it. We see that one and the same thing, namely black earthy bitumen that is often hard when it comes from the earth, is known by all of these names. It is necessary that I be able to defend my opinion with the words of the writers themselves. But first may I describe this kind of bitumen.

This bitumen is black, pitchy and resembles a poor quality of coal. It has the lustre of pitch and is sometimes soft, sometimes hard. Broken into splinters and set on fire it burns. When the soft variety is pulverized and placed in olive oil it dissolves rapidly, as Galen has written. I have purchased many of these stones, tabular and black and similar to the tablets that come from Coele-Syria, which, when placed in a fire, burn with a meagre flame. These come from the hills around the Dead Sea and from the same locality that produces bitumen. Their odor is similar to that of bitumen. However, without doubt the hot waters flowing deep beneath those bituminous hills have abraded and liquefied the bitumen and carried it into the lake or sea in which these waters have welled up as springs. This black bitumen is later cast up by the raging and seething of the sea and is found on the shores. It differs from the true Judean bitumen only in hardness. The more broken material is not tabular but differs only in this respect. Ampelitis earth, which is also called pharmacist's earth (pharmacitis) is the best. It is pitch black and when broken into small splinters has the same lustre as small pieces of coal. It dissolves when crushed and mixed with olive oil, according to Dioscorides. Galen says that it differs greatly from other earths and comes near to being the essence of stone. From this we know it to be hard. It is known to consist of bitumen, not so much from the writings of Pliny who says that it is very similar to bitumen as from the writings of Posidonius who says that it is bituminous. Since it is such a material it burns readily. From all of these facts it is evident that ampelitis earth is native bitumen of the finest quality and for that reason is given preference by physicians. Bitumen that has been mixed with too much earth is usually of another color and does not belong to this group.

It is obvious to the eyes that native coal does not differ from ampelitis earth. Theophrastus who has referred to coal makes no reference to ampelitis and, on the other hand, those who have written about ampelitis have made no reference to coal. Theophrastus writes that native coals are those earthy substances which we regard as varieties of stone and earth and which, for that reason, we call by a name formed from stone and coal. Certainly some are much harder than others. It is obvious that every writer should have given a reason for the selection and meaning of his name, for example, those who have called this native bitumen, earth, or have said that it is an earthy coal and those who call it stone or λιθώδια coal because they regard coal as being soft and this material as being hard. For the same reason some have called a similar hard bitumen jet because they classify it as a stone. The Germans call it by a name composed of a corruption of the original name to which they have added the name for stone. Since it actually belongs to this genus a description of it will suffice.

Jet, according to various writers, is black, tabular, quite light, inflammable and with an odor similar to bitumen. The most inflammable is the best. If sprinkled with water it will burn stronger but sprinkling with oil extinguishes the fire. Nicander writes that it is not consumed by fire either because bitumen burns continuously within the earth in many places or because the Magi, who use it in what they call divination, according to Pliny, deny that it is consumed if they so wish it for any reason. Serpents are driven away by its fumes. Having been warmed by rubbing it will pick up small light objects as does amber. It will not do this unless it has a natural brilliancy or has been polished. Solinus says that this property is limited to gemmy jet while Dioscorides, on the other hand, says that rough unpolished jet has the same property. Pliny writes that jet does not differ greatly from wood and since it is by nature tabular it has this type of fracture. It breaks with ease when dropped. The gem samothracia is seen to be nothing other than polished jet since Pliny writes that it has the same color, lightness and resemblance to wood. He writes, a black gem with a weight similar to that of wood is given the name of the island of Samothrace.

Lapis thracius is identical with jet. According to Euax Maurus it is black. According to Nicander and Dioscorides when it is sprinkled with water it burns with a very clear flame and is entirely consumed the same as bitumen and when sprinkled with oil it will not burn. An unknown Greek writer states that when burned it has an odor similar to that of bitumen but so acrid and unpleasant that no serpent will remain near it. In addition, a follower of Nicander correctly observes that it is bitumen because it has hardened after the manner of a stone. Lapis obsidianus is the same as jet since it is very black; because it gives a shadowy reflection similar to the image in a mirror; and because small statues have been made from it. In fact these and other similar criteria are those by which jet is distinguished from stones. Moreover we know this concerning obsidianus since Pliny writes as follows concerning glass. Obsidianus is classified as a glass because of a similarity to a stone found at Obsidius, Ethiopia, which is very black and sometimes transparent. It has a dense appearance and reflects a dark shadowy image like a mirror. They make many gems from it and the statue of Augustus was cut from the massive dense material as well as the four elephants in the temple of Concordia which was consecrated by a miracle. Tiberius Caesar, when placed over Egypt, sent back an obsidian statue of Menelaus that came into his possession, to the priests of Heliopolis. This shows the ancient origin of the material that is now falsified with glass. Xenocrates writes that lapis obsidianus occurs in India; Samnium, Italy; and near the ocean in Spain. Pliny, when writing about this mineral, says that it comes from India.

I shall mention the places where this bituminous earth occurs. It is mined in that part of Britain or Albion which we call Scotland because of the Scotch Germans who emigrated there. It is found most abundantly about twenty miles from Edinburg on the Deisert heath at a place they call Carbon. Some of this material burns as I have described elsewhere."

The rest of the book is taken up by the discussion of amber, anatachates, baptes, lipare and other favourite minerals of the authors of ancient and medieval lapidaries, which Agricola wrongly believes to belong to the bitumen family.

Andreas Libavius (Andrew of Libau, 1540-1616) attempted to carry out this method of classification by testing further to a logical conclusion. Libavius was a physician who during the latter part of his life was rector of the Latin school at Rothenburg. In his Singularium [1] we find an essay on "the Origin and Nature of Amber" in which he discusses the conclusions drawn by Agricola and by Andreas Aurifaber (Andrew Goldschmidt, who in 1541 wrote an essay on amber). This essay consists of ten chapters. In the first chapter Libavius gives the theories of the ancient poets and naturalists who believed that amber was formed from foam of the sea, the urine of a lynx, resins inspissated by the heat of the sun or simply from juices secreted by certain trees. In the second chapter he reviews the opinions of commentators on the classics who tend to regard amber as a resin transformed in the earth by the "bituminous heat", the mineral-building force propagated more particularly by Agricola and Aurifaber, or "boiled by the sea" as Cardanus states.

In his third chapter Libavius says that he cannot agree with such views for amber contains fossils, the remains of animals and plants, and hence it cannot have been formed recently from resins exuded by trees in the northern regions like Sweden or Finland. In the next two chapters he argues that amber may be a fossil resin but he has no proofs that resinous trees grew in the north in such remote periods. Neither could it have been formed from minerals which some authors claim to exude fatty or resinous substances. A review of the different types of amber in various countries show that there is no such connection between this substance and specific minerals. Nor is the salt sometimes found in amber a proof of maritime origin of amber, for other specimens do not contain any salt on incineration.

After reviewing the classical evidence on such substances as gagates, obsidian, black resin, pharmacitis and allied substances Libavius concludes that one cannot argue on the classical written evidence alone and should test such substances for proper classification. If we arrange the members of the bitumen family in order of increasing consistency we get the series: naphtha, bitumen liquidum (maltha), asphalt, pissasphaltos, gagates, lithatrance, and pharmacitis. Amber (karabé, electrum or glessum), he says, is identical with gagates and therefore a member of the bitumen family. Gagates differs slightly in colour and odour but when distilled it yields first a milky, watery distillate, then an inflammable oil, a liquid bitumen, a thick distillate which congeals to a "pissasphaltos" and the residue (caput mortuum) is a black bituminous substance. If we distill amber we also get a milky distillate, a sulphurous distillate,

---

[1] Andreas Libavius, D.O.M.A. Singularium Andreae Libavii, Med. Phys. Rotemb. Pars Prima (Peter Kopff, Frankfort, 1599, pp. 208-251).

a clear, water-white "petroleum", a yellowish oil, a heavier bitumen liquid and finally a black residuum (feces) which is bituminous. Hence his conclusion is that amber is a kind of gagates, a member of the bitumen family. His tests show that he has personally analysed these bitumens with the crude methods at his disposal and has tried to sort out their individual and their group characteristics. As to the salt sometimes found on the incineration of such members of the bitumen family, he points to the bitumen of the Dead Sea and argues that such salts are intrusive compounds which will contaminate bitumens in contact with salt waters.

Libavius attempts certainly to carry sixteenth century classifications of the bitumens to their logical conclusion. With the crude methods at their disposal and the limited variety of samples nothing better could be achieved at the time. This changes when, during the seventeenth century, new seepages of crude oil and new types of bitumens were discovered both in the East and in the New World [1]. In view of the fact that bitumen and petroleum had but few applications outside medicine only relatively small quantities were imported. Therefore, if new data on such crudes and bitumens were now available, samples of such relatively expensive imports were seldom put to test. Unfortunately we still have few data on the quantities of petroleum shipped to Europe from these distant countries. A search for data on the trade in Barbados Tar, Rangoon Tar or petroleum from the Persian Gulf in the papers of the East India Company and similar trade associations will probably reveal many interesting facts.

Such data exist on the trade in crude from the seepages of North Sumatra (Atchin), which crude "oly van aerde" (Malay: minyak tanah) was exported by the Dutch East India Company to Amsterdam and then sold even to Europe's largest producer of petroleum, Italy, where this "minyak tanah" managed to find its way into the official pharmacopoeia. When in 1632 captain Stadtlander was sent to Atchin as an ambassador [2] he was ordered to buy "three or four pots of "oly van aerde", called minyak tanah by the Atchinese". In 1636 Jacob Compostel receives the same instructions and the Sultan of Atchin promises him two jars. Compostel writes home "I have not been able to get more than two pots of petroleum but hope to meet your demand for five to six jars soon (Letter of December 22, 1636). On July 29, 1642 Peter Soury reports "Today I managed to get the four requested Siamese jars of pure petroleum which cost $3\frac{3}{4}$ "realen van achten" (about £ 3.15.—). A letter of September 16, 1655 informs Amsterdam "herewith two Siamese jars

---

[1] R. J. Forbes, Studies in Early Petroleum History I (Leiden, 1958, p. 141).

[2] J. Ligtvoet, Aardolie in Atjeh in de zeventiende eeuw (Bijdragen Taal-, Land- en Volkenkunde Ned. Indië, 4e reeks, Deel II, 1878, pp. 378 ff.).

containing 33 bamboos (a liquid measure) of petroleum, together with the pottery jars and packing a total of twenty guilders and eight pence". Then the quantities slowly grow. In 1667 twelve different ships take back to Amsterdam 380 pounds of oil with a total value of 570 guilders (that is 30 pence a pound). In 1668 902 jugs (at twopence a jug) and 61 jars (total value 90 guilders four pence) went home, in 1669 75 jars of "petroleum from Pegir" (of 36 litres each) of a total value of 252 guilders reached Amsterdam.

Hence most books discussing bitumens simply print a rehash of the theories and facts from Agricola and Libavius, often in a very uncritical fashion. To quote just one example, this is what Giambattista della Porta says in his very popular Magia Naturalis, originally published at Naples in 1558; when discussing "compositions with burning waters" in his twelfth book on "artificial fires"[1]:

'The first kind (of bitumen) is liquid, called Naphtha, we call it Oyl of Peter, which remains in stones and Kisram. This has great affinity with Fire, and the fire will take hold of it every way at a great distance. So some say, that Medea burnt a whore, who, when she came to sacrifice at the altar, the fire laid hold of her garland. Another kinde is, that men call Maltha; for in the city of Camegens Samosata, there is a lake sends forth burning mud: when any solid thing thoucheth it, it will stick to it; and being touched it will follow him that runs from it. So they defended the Walls, when Lucullus besieged them, and the Soldier burned in his Armor. Waters do kindle it, and only Earth can quench it, as experience shows. Camphire is a kinde of it: as Bitumen, it draws fire to it, and burns. Pissasphaltum is harder than Bitumen: both Amber and Jet are of this sort; but these burn more gently, and not so much in the waters. Moreover, in regard it burns in the Water, it is Brimstone; for no fatter thing is dug forth of the Earth. To maintain this fire, itself is sufficient: it neither burns in the waters, nor is it put out with water, nor does it last long; but, joyned with Bitumen, the fire will last always, as we see in the Phlogrean Mountains at Puteoli: and as fire, if Oyl be cast in, burns the more; so when Bitumen is kindled, water cast on, makes the flame greater."

If, therefore, we turn to the first encyclopedias published early in the eighteenth century [2] we cannot expect to find more useful information. Thus the famous Chambers' Cyclopaedia contains the following entries [3]:

"ASPHALTOS, or ASPHALTUM, a solid, brittle, black, inflammable, bituminous substance, resembling pitch, brought from the east, and particularly Judea; whence it is also called Jews-pitch.

---

[1] John Baptista Porta, Natural Magick (Basic Books Inc., New York, 1957) (A reprint of the original English edition of 1658 printed by Thomas Young and Samuel Speed at London).
[2] J. Harris, Lexicon Technicum (London, 1704).
[3] E. Chambers, Cyclopaedia (4th edit. London, 1742, 2 vols).

The asphaltos of the Greeks is the bitumen of the Latins. Modern naturalists, who make a class of bitumens, place asphaltos at the head of it; as being the farthest maturated and concocted of the whole tribe; but consisting of the same simple principles as the rest. It is chiefly found swimming on the surface of the lacus asphaltites, or dead sea, where anciently stood the cities of Sodom and Gomorrah. — It is cast up from time to time, in the nature of a liquid pitch, from the earth which lies under this sea; and being thrown upon the water, swims like other fat bodies, and condenses by little and little, through the heat of the sun, and the salt that is in it: it burns with great vehemence; in which it resembles naphtha, but is thicker as to consistence.

The Arabs use it to pitch their ships withal, as we do common pitch. — Besides, there was a deal of it employed in the embalming of the ancients.

It is supposed to fortify and resist putrifaction; resolve, attenuate, cleanse, and cicatrize wounds: but is little used among us either externally or internally. It is usual to sophisticate the asphaltos, by mixing common pitch along with it; the result whereof makes the pissasphaltum, which the coarseness of the black colour, and the fetid smell easily discover. — Others, however, will have its pitchy quality natural to it, and suppose pissasphaltum to be the native asphaltum.

ASPHALTUM also denotes a kind of bituminous stone, found near the ancient Babylon, and lately in the province of Neufchâtel; which, mixed with other matters, makes an excellent cement, incorruptible by air, and impenetrable by water; supposed to be the mortar so much celebrated among the ancients, wherewith the walls of Babylon were laid.

It yields an oil which defends ships from water, worms, etc. much better than the ordinary composition; and which is also of good service for the cleansing and healing of ulcers, etc.

BITUMEN, in a general sense, a fatty, tenacious, mineral juice, very inflammable: or a fossil body which readily takes fire, yields an oil, and is soluble in water.

Naturalists distinguish three kinds of bitumens, hard, soft, and liquid or oily; each of which they subdivide into several others.

Among the hard bitumens are ranked yellow amber, sometimes amber-grease, jet, asphaltum, or Jews-pitch, pissasphaltum, pit-coal, black-stone, and sulphurs. — The soft are maltha, bitumen of Colao, of Surinam, and Copal. — Lastly, the naphta of Italy, and petroleum are ranked among the liquid bitumens, to which may be added zacinthus.

Of bitumens some again are fossil, others are found floating on the surface of certain lakes, and others spring from the earth like fountains: as at Pitchford in Shropshire, etc. — Some bitumens are so hard, that they are used in forges, instead of coals: others so glutinous, that they serve instead of cement, or mortar in buildings; of which kind it was, that the famous walls of Babylon were built: and others so liquid, that they are burnt in lamps instead of oil.

The bitumen in most esteem is that of Judaea.

NAPHTHA [1], a kind of liquid bitumen, very oily, and inflammable; exuding out of the earth in several places in Chaldaea; particularly the place where stood the

[1] The word, in the original Chaldee, signifies stillare, to ooze, or drop; naphtha, according to Pliny, running like a kind of bitumen.

ancient Babylon: and found also in some provinces of Italy and France, particularly Auvergne, and near Ragusa.

Naphtha is found swimming on the surface of the water of some springs. It is usually of a black colour, though that found in certain springs about Babylon is said to be whitish. That of France is soft and black, like liquid pitch, and of a fetid smell; that of Italy is a kind of petrol, or a clear oil, of various colours, oozing out of a rock, situated on a mountain in the dutchy of Modena.

Naphtha is esteemed penetrating, resolutive, and vulnerary; but its virtues are little known in medicine: its chief use is in lamps, etc. on account of its inflammability.

The Turks call the naphtha, cara sakiz, black mastic, to distinguish it from pitch. Vossius has an express treatise on naphtha, ancient and modern: He says, it is a flower of bitumen, of more virtue than any other bitumen.

PETROL, PETROLEUM, q.d. petrae-oleum, oil of petre, or rock oil, an oleaginous juice, supposed to issue out of the clefts of rocks, and found floating on the waters of certain springs.

Beside artificial and vegetable oils, *i.e.* those drawn from plants, etc. by expression, there are also natural and mineral oils issuing of themselves from the entrails of the earth, called by a common name petrols, or petrolea.

These, according to all appearance, must be the work of subterraneous fires, which raise, or sublime the more subtile parts of certain bituminous matters that lie in their way. These parts being condensed into a liquor by the cold of the vaults of rocks, are there collected, and ooze thence through clefts and apertures, which the disposition of the ground furnishes them withal. Petrol then is a black liquid bitumen, only differing by its liquidity from other bitumens, as asphaltum, jet, etc. See BITUMEN.

The naphtha, which is either a liquid, or at least a very soft bitumen, is much the same with petrol. Hitherto there has been little petrol found, except in hot countries. Olearius says, he saw above thirty springs of it near Scamachia in Persia: there are also petrols in the southern provinces of France; but the best are those in the duchy of Modena, first discovered by Ariosto a physician in 1460, in a very barren valley, twelve leagues from the city of Modena. There are three canals dug with great expence in the rock; by which three different kinds of petrol are discharged into little basins or reservoirs; the first as white, clear, and fluid as water, of a brisk penetrating smell, and not disagreeable; the second, of a bright yellow, less fluid, and a less brisk smell than the white; the third, a blackish red, of a thicker consistence, and a smell more approaching that of bitumen.

M. Boulduc has made several experiments on the petrol, described in the Hist. of Acad. of Scienc. an. MDCCXV. He observes, that he could not raise from it any phlegm or saline spirit by any distillation, either in balneo mariae, or in a sand heat: all that would rise was oil; at the bottom of the pelican remained and exceeding small quantity of a thickish, brownish matter. Hence, to use petroleum in medicine, it must be prescribed just as it is. It is a remedy nature has prepared to our hands; it is found very warm and penetrating, and commended in many outward complaints, rheumatick and arthritick pains and paralytick limbs.

PISSASPHALTUM [1], or PISSASPHALTUS, in natural history, denotes a native, solid

---

[1] The word is compounded of πισσα, pitch, and ασφαλτος, bitumen.

bitumen; found in the Ceraunian mountains of Apollonia: of an intermediate nature between pitch and asphaltum.

PISSASPHALTUM is also a name given to a factitious substance compounded of pitch, and asphaltus or bitumen Judaicum.

The coarseness of the black colour, and the fetidness of the smell distinguishes it from the true asphaltum. PISSASPHALTUM is also used by some writers to denote the Jewish pitch, or simple asphaltum."

Even the excellent French pharmacists like Baumé and Macquer who did so much to transform qualitative chemistry into quantitative chemistry had little to contribute. Macquer [1] who uses the word "bitumen" in the sense of "combustible mineral" like Agricola defines petroleum as a kind of liquid bitumen and has nothing more to say on the subject. The reasons for this lack of real knowledge are clear.

In the eighteenth century knowledge of petroleum and related substances had not yet advanced very much, mostly due to the lack of proper analytical tools and an overdose of theory little related to accurate observation. Even the best like Homer sometimes nodded and this becomes clear when we turn the pages of the chemical manual by Herman Boerhaave (1688-1738) [2] which was widely read along with his medical writings. Though he certainly sponsored proper scientific observation we read in his Elements of Chemistry:

"One must also include in the sulphur category those fatty substances which contribute most towards its composition and which are natural products of the earth. Such a substance is petroleum, or rock oil. Its name is enough to explain its nature and origin. It is pressed out of melted bitumen; it issues forth from rocks; it is very subtle; very light; strong smelling and extremely inflammable. It often floats on the water of a spring. Most of its properties are so similar to those of distilled oil that many people believe it was made by a subterranean fire. It is also frequently called a liquid bitumen from which, however, it differs a great deal in colour, odour and clarity.

Naphta is very similar to petroleum, but more watery, thinner and clearer. It is extremely inflammable and once it has caught fire it stays alight a long time and is difficult to extinguish. It is the purest and most subtle part of bitumen, — the flower thereof.

The bitumen of the Latins, which is the asphalt of the Greeks, is thicker than naphtha or petroleum. It is a very tough substance but is still more or less fluid when in its primary form: when in its natural state it floats on the surface of water: it catches fire very quickly.

---

[1] P. J. Macquer, Dictionnaire de Chimie (Paris, 1778).
[2] Herman Boerhaave, Elemens de Chymie (trad. J. N. S. Allemand, Amsterdam, 1752, pp. 57-59; 367-370; 432-433). The first original edition of the Elementa Chemiae was published at Leiden in 1732 in 2 vols; another edition in the same year at Leipzig (Caspar Fritsch) and an English edition (abridged) was published by J. Wilford (London) in 1732 too, the second French edition at Leiden in 1752.

This bitumen, baked and dried by the sun, the force of fire or the passage of time, becomes glossy, heavy and harder than pitch: it can be melted again in the fire; it mixes well with other oils; it is inflammable. It is what is known as Jew's pitch or bitumen.

Pissasphalt

Pissasphalt, as its name indicates, is a substance midway between pitch and bitumen: it is a black substance, earthy with a strong and unpleasant smell and it only seems to differ from the above-mentioned substances in degree. Hence it is possibly only a product of art or nature made up of various fatty matters with liquid bitumen added.

Jet

When nature has so far perfected it that it becomes black, hard, earthy, brittle, polished, strong-smelling and glossy it appears that it forms that bituminous stone known as jet or lapis Thracius Nicandri.

Stone carbon

When the fatty parts of bitumen mix with and coagulate with lumps of stony soil or metallic slag they form a hard substance composed of various sheets or layers, which is black and greasy, brittle and inflammable. This is fossil carbon or stone carbon which also belongs here.

Succin

That which is called amber, carabé, succin or electrum also belongs to this group, for it would appear that it owes its origin to a bituminous sulphur; it is inflammable and melts in fire. It is composed of an acid salt which is as much liquid as solid, and of a fossil oil very similar to petroleum. There are different kinds of it—white, lemon-coloured, yellow, black and red.

Earth oil

It is only very rarely that we obtain from the Indies earth oil described by Neuhovius. The princes of Asia keep it for themselves so I am not able to determine whether it is a kind of petroleum or naphtha.

As for that which is brought to Europe and sold in the shops under that name, a person well versed in this sort of thing told me that it is made with oil pressed from coconuts and mixed with some medicinal earth; it therefore belongs entirely to the vegetable category. And is not the oil which is called Barbados oil made in the same way?"

In a further passage he discusses the problem of why certain minerals are combustible:

"*Fuels obtained from the Fossil Kingdom*

One of the first things which it is important to notice here is that the same law of combustibility which holds good for the vegetable and animal categories also applies to fossils, for it will be seen that in the latter only the oils are inflammable and that the other components are not. Even their various types of oil produce less smoke, soot and cinders according as they are more subtle and lighter; and conversely the thicker and heavier they are the more they produce. These oils may occasionally be of a subtlety approaching that of alcohol, although I do not know of any kind found up to the present which is subtle enough to mix with water.

Naphtha is the most similar to alcohol

I have certainly read that in some places a certain liquid, which was distilled from rocks, caught fire when a lighted candle was brought near to it, and I remember, too, that a liquid has sometimes been observed issuing from certain springs which caught fire in the same way. But those to whom we owe these observations did not inform us whether these combustible liquids had also the property of being miscible with water. There are historians who tell us that the naphtha of Babylon was so subtle and volatile that it caught fire with great ease and produced so safe a flame that if it were spread about in the streets it would catch fire from the flames of the torches that were carried at night, so that you would have said that it had caught fire of its own accord. Whole stretches of the street were seen to be dotted with a blue flame, although it was a feeble one which did hardly any harm. This makes me suspect that this liquid had a subtlety very close to that of alcohol, for perhaps in hot countries our alcohol, sprinkled about in the same way, would catch fire in the way we have previously seen when it was evaporated under a bell jar and a match brought near it. But as it is hardly possible to obtain this true naphtha at any price, nothing certain can be said about it yet. That which is sold to us in Europe by this name is far removed from this high inflammability. It is much thicker and tougher.

After naphtha petroleum is the most similar

Petroleum is, in fact, also a subtle liquid, but is not comparable, however, to the naphtha of the ancients or to our alcohol. When it is rectified by distillation it is always rendered more subtle and inflammable, but it always remains oil and does not become alcohol. Moreover the same thing happens in this case as with vegetable oils: *i.e.* the purer, subtler and lighter is the oily, inflammable matter found in fossils, the less it produces smoke, soot, an evil smell and cinders, and at the same time its flame is lighter, purer and feebler.

Stone carbon

The other inflammable fossils wherein is mixed a greasy, heavy, incombustible matter catch fire with much more difficulty. They have to be exposed to the action of the wind or a bellows in order to burn strongly, but they also produce a flame and fire which is all the more violent: it can be seen very clearly in stone carbon when it is on fire. These materials also give off very thick, black smoke even a little malodorous, especially when it is condensed into soot. They also leave a large amount of coagulated cinders, usually insipid, but very heavy."

Only the "oily part" of natural asphalts and bitumes is combustible. He then goes on to discuss sulphur and its combustion (oxidation) products, but as he still believes that combustion is possible without air his conclusion can only be that "fire rarifies all bodies". In his theory of chemistry, which takes up the second part of this French edition he gives a few examples of such combustion with the exclusion of air in these words:

'Before ending this article I will add the following remarks relating to the nature of fire. Fire does not need air, nitre, fuel, sulphur or any other matter in order to exist. The true naphtha is of all known materials that which catches fire most easily. It even catches alight at quite a distance from the flame, like pure petroleum (Journal des Scavans. 1675 p. 53). Bodies rubbed with naphtha, once on fire, continue to burn when they are immersed in water (Journal des Scavans. 1683 p. 104). Naphtha is lit by the flame of a candle enclosed in a lantern and which in consequence it does not touch (Transact. Philos. N. 100 p. 188).''

However, by the end of the eighteenth century Lavoisier, Black, Priestley and their generation rapidly transformed qualitative alchemy into quantitative chemistry. Now the tools were being forged for a rational and logical analysis of petroleum and theories were conceived which really led to proper understanding of the compounds which made up that intricate mixture of hydrocarbons which we call petroleum.

Lavoisier himself had subjected various compounds to combustion and had analysed the combustion products, thus he was able to calculate the percentual composition of his original sample. Berzelius, Gay-Lussac, Prout and von Liebig succeeded in devizing simple yet accurate methods for this "elementary analysis" of organic compounds, which method was ready in essence about 1835 and which has only been perfected in details since. This elementary analysis helped to establish that petroleum, naphtha and similar substances were compounds of carbon and hydrogen, sometimes containing small percentages of nitrogen, oxygen and sulphur.

The earlier eighteenth-century chemists had no such simple tools at their disposal, they resorted to what we now know to be destructive distillation of crude oils and men such as Winterl or Martinovich, who examined Galician crudes [1] could only conclude that crude oil consisted of a "buttery clear oil", some salt, gases and a carbonaceous residue. De Saussure [2] equipped with better tools established in 1817 that an Italian crude consisted of hydrocarbons. His results were confirmed by Döbereiner and by Dumas [3], the latter concluded that the ratio of carbon to hydrogen was about $n : 2n + 2$. This was common knowledge about 1840, and in certain bitumens and crudes an appreciable amount of oxygen had also been found [4].

[1] Winterl, Crell's Chem. Ann. vol. I, 1788, p. 493.
  Martinovich, Crell's Chem. Ann. vol. I, 1788, pp. 32 & 162.
[2] H. de Saussure, Ann. Chim. Phys. vol. 4, 1817, p. 314.
[3] J. B. Dumas, Liebigs Ann. vol. VI, 1833, p. 257; Blanchet and Sell, Liebigs Ann. vol. 6, 1833, pp. 308 ff.; Pelletier & Walter, J. de Pharmacie vol. 26, 1833, pp. 549 ff.
[4] J. P. Boussingault, Ann. Chim. Phys. vol. 64, 1837, pp. 141-151.

It was soon appreciated that petroleum and similar substances formed part of a special domain of chemistry, which Berzelius in 1814 had baptized "organic chemistry". However, the early nineteenth-century chemists were still a long way from appreciating the importance of recognizing, as we do, different series of hydrocarbons such as paraffins, aromatics, unsaturated and cyclic hydrocarbons. By 1825 they had realized that isomers and polymers of the same percentual formula existed side by side in nature. Slowly they realized that certain constituent atom-groups, which were called "radicals" represented a group of characteristic properties persisting in many derivatives of a certain compound. Work on such substances as coal-tar and wood-tar, which can be more easily split into a series of pure compounds, led to the recognition of "alcohols", "acids", "phenols" and "ketones". By 1845 the chemists started to characterize each organic compound carefully by determining its physical properties, such as its boiling point, its density and its melting or congealing point. However, there was no consensus yet on the actual grouping of the atoms within the molecule nor on the number of atoms in each specific molecule. In 1840 Dumas and Stas [1] had determined the atomic weight of carbon properly and had found it to be twelve, and Kékulé had found that the valency of the carbon atom was four [2], but only when Stanislao Cannizzaro (1826—1910) revived the half-forgotten hypothesis of his countryman, Amadeo Avogadro, that equal volumes of gases and vapours contain equal numbers of molecules, and published this in his Brief Sketch of a Course of Chemical Philosophy (1858) was a solution found for the conflicting views.

In 1859 Kékulé gave graphic formulae of organic compounds in his handbook of chemistry and two years later Butlerow discussed "structural formulae" [3]. Progress was now very rapid for the advantage of this newly-won theory was obvious. In 1865 Kékulé published [4] his views on the structure of benzole, ring structures and side chains of the aromatic compounds. By 1868 A. W. Hofmann had created the proper analytical tool, an accurate vapour-density test [5]. As yet this had no great impact on the knowledge of the composition of petroleum for most of this knowledge had been gained by analysing coal-tar compounds. When in 1820 Dr. Thomson distilled a "perfectly colourless naphtha from Persia" [6], he concluded that it contained "thirteen atoms carbon against 14 atoms hydrogen. A deficiency of 3%, he was

---

[1] J. Dumas and J. S. Stas, Comptes Rendus vol. 11, 1840, p. 991.
[2] A. Kékulé, Liebigs Ann. vol. 104, 1857, p. 129.
[3] A. Butlerow, Z. der Chemie vol. 4, 1861, p. 459.
[4] A. Kékulé, Liebigs Ann. vol. 137, 1866, pp. 129 ff.
[5] Hoffmann, Berichte vol. I, 1868, pp. 198 ff.
[6] Thomson, Properties of Native Naphtha (Quart. J. Science, Lit. Arts, vol. IX, 1820, p. 408).

inclined to think, was nitrogen". What the petroleum chemist knew of his base material was still mainly practical experience added to the very general fact that the ratio carbon to hydrogen tended to change a little from oil to oil.

A fair picture of this empirical European knowledge of the composition of petroleum about the time of Drake's discovery can be gained from Ure's dictionary. Andrew Ure (1778-1857) was professor of chemistry (1804-1830) at the Andersonian Institution (later Royal Technical College) at Glasgow and in 1844 he published the first edition of his Dictionary of Arts, Manufactures and Mines, which enjoyed a great popularity. Only a few years after his death, the 1860 edition was published [1] which under the headings "Bitumen" and "Naphtha" give very full information on our subject. Under the first heading Ure gives his views on the different branches of the bitumen family in these words:

"BITUMEN, or ASPHALTUM (Bitume, Fr.; Erdpech, Germ.) — A black substance found in the earth, externally not dissimilar to coal. It is composed of carbon, hydrogen, and oxygen, like organic bodies; but its origin is uncertain. It has seldom been observed among the primitive or older strata, but abundantly in the secondary and alluvial formations.

Bitumen comprises several distinct varieties, of which the two most important are asphaltum and naphtha.

Asphaltum is solid, and of a black, or brownish-black, colour, with a conchoidal brilliant fracture.

Naphtha — Liquid and colourless when pure, with a bituminous odour.

There are also the *earthy*, or *slaggy mineral pitch*—*petroleum*—a dark coloured fluid variety, containing much naphtha, and *maltha*, or *mineral tar*.

Springs of which the waters contain a mixture of petroleum, and the various minerals allied to it — as bitumen, asphaltum, and pitch—are very numerous, and are, in many cases, undoubtedly connected with subterranean heat, which sublimes the more subtle parts of the bituminous matters contained in rocks. Many springs in the territory of Modena and Parma, in Italy, produce petroleum in abundance; but the most powerful perhaps yet known are those of Irawadi, in the Burman empire. In one locality there are said to be 520 wells, which yield annually 400,000 hogsheads of petroleum.

Fluid bitumen is seen to ooze from the bottom of the sea on both sides of the island of Trinidad, and to rise up to the surface of the water. It is stated that, about seventy years ago, a spot of land on the western side of Trinidad, nearly half-way between the capital and an Indian village, sank suddenly, and was immediately replaced by a small lake of pitch. In this way, probably, was formed the celebrated Great Pitch Lake."

He then quotes Sir Charles Lyell, Manross and the Earl of Dundonald on the excellency of the bitumen from the Pitch Lake, and continues to give some details on other deposits of natural asphalts:

[1] Andrew Ure, Dictionary of Arts, Manufactures and Mines (Edinburgh, 1860; "Bitumen" vol. I,; "Naphtha", vol. III, pp. 220-233).

"Asphaltum is abundant on the shores of the Dead Sea. It occurs in some of the mines of Derbyshire, and has been found in granite, with quartz and fluor spar, at Poldice, in Cornwall. There is a remarkable bituminous lime and sandstone of the region of Bechelbronn and Lobsann, in Alsace. From the observations of Daubrée, we learn that probably this bitumen has had its origin as an emanation from the interior of the earth; and indeed, in Alsace, with the great elevated fissure of the sandstone of the Vosges, a fissure which was certainly open before the deposit of the Trias, but was not yet closed during the tertiary epoch, affording during this latter, moreover, an opportunity for the deposition of spathic iron ore, iron pyrites, and heavy spar."

Next we get some details on the "Elastic Bitumen, called also mineral caoutchouc and elaterite" which was first observed in Derbyshire "in the forsaken lead mine of Odin, by Dr. Lister, in 1673, who called it a subterranean fungus. It was afterwards described by Hatchett"[1]. He compares it with the "French elastic bitumen" discovered at the coal mines of Montrelais (Auvergne) by giving the percentages of carbon, hydrogen, nitrogen and oxygen, which differ little.

The long entry on "Naptha" was written by C. Greville Williams (1829-1910), a consulting and industrial chemist, who in the same year 1860 discovered isoprene by destructive destillation cf natural rubber. Indeed many of his "naphthas" are simply products of destructive destillation (cracking) of natural products which he groups together:

"NAPHTHA. By the term naphtha, we understand the inflammable fluids produced during the destructive distillation of organic substances. Formerly the term was confined to the fluid hydrocarbons, which issue from the earth in certain parts of the world, and appear to be produced by the action of a moderate heat on coals or bitumens. The term has now, however, become so extended as to include most inflammable fluids (except perhaps turpentine) obtained by distillation from organic matters. We shall study the various naphthas under the following heads:

Naphtha (Boghead or Bathgate)
„ (Bone or Bone Oil)
„ (Caoutchouc or Caoutchoucine)
„ (Coal)
„ (Native)
„ (Shale)
„ (Wood)."

The method of manufacturing the "Boghead or Bathgate Naphtha" will be discussed in a subsequent chapter on the story of paraffin wax but we should study here Williams' method of analysing his products for this pertains to other types of naphtha as well:

[1] Charles Hatchett, Observations on Bituminous Substances with a description of the Elastic Bitumen (Trans. Linn. Society vol. IV, 1798, pp. 129-154).

"*On the chemical nature of the fluid hydrocarbons constituting Boghead naphtha.*

It has been said, in the above condensed account of the process for preparing paraffine oil from coal, that when the crude oil is rectified with water, a clear transparent naphtha is obtained. This fluid, as found in commerce, is by no means of constant quality. By quality, we mean the power of distilling between given limits of temperature. Some kinds are of about the same degree of volatility as commercial benzole, while others distil at nearly the same temperatures as common coal naphtha. The hydrometer is not a safe guide in choosing this naphtha; this arises from the fact that photogens, of very different degrees of volatility, have almost the same densities. The safest plan is to put the fluid into a retort, having a thermometer in the tubulature, and distil the contents almost to dryness. The careful observation of the range of the mercurial column during the operation is the best mode of ascertaining the quality of the fluid.

The more volatile portions which distil over with water, are free from solid bodies, and consist of a mixture of fluids belonging to three series of homologous hydrocarbons, namely,

The benzole series;
The olefiant gas or CnHn series; and
The radicals of the alcohols."

Williams is well aware of the complex composition of such naphtha and he carefully describes the "proximate separation of complex mixtures of hydrocarbons" by preparing "ten degree fractions" by careful distillation to obtain constant boiling points if possible. "The more volatile portions may be tested for benzole by converting them into aniline. The simplest way of detecting the CnH_{2n} series (homologous with olefiant gas) will be by ascertaining whether the naphtha is capable of decolorizing weak bromine water". After bromination and drying the oil is distilled "when the radical and benzole series of hydrocarbons (our "paraffins" and "aromatics") will distill away, leaving the brominated oil, which may then be distilled in a vessel by itself". Separating the "radical" and the "benzole" series is achieved by treatment with nitric acid in small portions. "The hydrocarbons, unacted on, rise to the surface in the form of a transparant brilliant fluid".

"When the treatment with acid has been repeated a sufficient number of times, the fluid is to be placed in a clean flask and well agitated with a solution of caustic potash, which will remove the nitrous vapours which are the cause of the green colour. The purified hydrocarbon is then to be separated by a tap funnel from the water, and dried by digestion with sticks of caustic potash. If it be desired to obtain the radical in a state of absolute purity, it must be distilled three or four times over metallic sodium.

The indifferent hydrocarbons obtained by the above process are colourless mobile fluids, having an odour somewhat resembling the flowers of the white thorn. They are very volatile, even at low temperatures, and have an average density of about 0.716. When the fractions with proper boiling points have been selected, it will be

THE CHEMISTS AND THE COMPOSITION OF PETROLEUM

found that they correspond in specific gravity, percentage composition, and vapour density with the radicals of the alcohols, as will appear by the following table, where the experimental results obtained by the author of this article in his examination of Boghead naphtha, are compared with the numbers found by other observers with the radicals obtained by treatment of the hydriodic ethers by sodium, and also by the electrolysis of the fatty acids.

Comparative Table of the Physical Properties of the Alcohol Radicals, as obtained from Boghead Naphtha, with those procured from other sources

Radicals	Formulae	Boiling Points, Fahr.				
		Frankland	Kolbe	Wurtz	Brazier and Gosslett	C. G. Williams
Propyle	$C^{12}H^{14}$					154.4°
Butyle	$C^{16}H^{18}$	311°	226.4°	222.4°		246.2
Amyle	$C^{20}H^{22}$			316.4		318.2
Caproyle	$C^{24}H^{26}$			395.6	395.6°	395.6

Radicals	Formulae	Densities				Vapour Densities			
		Frankland	Kolbe at 64.4°	Wurtz at 32°	C. G. Williams at 64.4°	Frankland at 51.8°	Kolbe	Wurtz	C. G. Williams	Theory
Propyle	$C^{12}H^{14}$				0.6745				2.96	2.97
Butyle	$C^{16}H^{18}$	4.899	0.6940	0.7057	0.6945	0.7704	4.053	4.070	3.88	3.94
Amyle	$C^{20}H^{22}$			0.7423	0.7365			4.956	4.93	4.91
Caproyle	$C^{24}H^{26}$			0.7574	0.7568			5.983	5.83	5.87

The acids and bases accompanying the hydrocarbons in Boghead naphtha have not yet been fully investigated; it has, however, been ascertained that certain members of the phenole series of acids and pyridine class of bases are always present. The quantities present in the naphtha of commerce are small in consequence of the purification of the fluid by the agency of oil of vitriol, followed by a treatment with caustic soda."

We are then informed about:

"NAPHTHA, Bone. Syn. Bone Oil; Dippel's Animal Oil. This fluid is procured in large quantities during the operation of distilling bones for the preparation of animal charcoal. The hydrocarbons of bone oil have not as yet been examined, but it has been found that the benzole series are present, accompanied by large quantities of basic oils. The acid portions are also uninvestigated. The bases have been very fully studied by Dr. Anderson, who discovered in bone oil the presence of no less than ten bases, several of them being quite new".

But apart from the nitrogen bases on which Williams gives further details he does not expect "Bone Naphtha" to be of much importance in the future, even if it had figured as "Oleum animale dipelli" in many older handbooks on pharmacy and chemistry.

Williams can of course not refrain from giving a few lines on "Naphtha from Caoutchouc, Caoutchoucine, Caoutchine", the average analysis of which would point to "hydrocarbons of the formula $n(C_5H_4)$. It is one of the best solvents known for india-rubber".

The type of naphtha discussed next is much more important:

"NAPHTHA, Coal. Ordinary coal naphtha is procured by the distillation of coal tar. The latter is placed in large iron stills, holding from 800 to 1500 gallons, and distilled by direct steam. As soon as the specific gravity of the distillate rises to 0.910, the naphtha is pumped into another still, and distilled with direct steam until the distillate again becomes of the density 0.910. It then constitutes what is termed "rough naphtha".

The residue obtained in the first distillation is run off into cisterns or tar ponds to allow of the removal of the water. This residue is called boiled tar. Pitch oil may be obtained from it by distillation with the naked fire, every 1000 gallons will yield about 320 gallons of pitch oil. The residue of pitch in the still is run out while in a melted state. The rough coal naphtha contains a great number of impurities of various kinds; the principal cause of the foul odour being the organic bases described in the article NAPHTHA, Bone. To remove these the naphtha is transferred to large cylindrical vessels lined with lead. These vessels contain a vertical axis passing down them, supporting blades of wood covered with lead, and pierced with holes. The axis or shaft has, at its upper end, a crank to enable it to be rotated. The naphtha having been run into the vessel, sulphuric acid is added, and the shaft with its blades made to revolve. By this means the naphtha and acid are brought into intimate contact. The whole is then allowed to settle, and the vitriol which has absorbed most of the impurities, and acquired, in consequence, a thick tarry consistence, is run off. This acid treacly matter is known in the works as "sludge". The naphtha floating above the sludge is then treated a second time with acid, if the naphtha be required of good quality. During the process, the naphtha acquires a sharp smell of sulphurous acid, and retains a certain amount of sulphuric acid in solution. The next process is to treat it with solution of caustic soda to remove these impurities. This may be effected in an apparatus similar to the first. The naphtha, after removal of the caustic liquor, is next run off into a still, and rectified; it then forms the coal naphtha of commerce. The ordinary naphtha of commerce is often very impure, owing to insufficient treatment with oil of vitriol. The author of this article has obtained from one gallon of commercial naphtha, as much as one and a half ounces of the intensely odorous picoline, mixed with certain quantities of other bases of the same series, and also traces of aniline."

Williams then points out that coal-tar naphtha contains a large percentage of nitrogen bases like pyridine, chinoline, aniline and pyrrol as well as aromatics such as benzole, toluole, xylole and cumole. The higher fractions of coal tar have not yet been investigated. All these hydrocarbons are separated by careful distillations and re-distillations, and Williams describes several stills used for this purpose in detail. He also discusses the properties of naphthalene and of the "pyrène" and "chrysène" prepared from coal-tar by Laurent [1].

Finally we come to the sections on "Native naphtha" and "Shale naphtha" which we will quote in full:

"NAPHTHA, Native. In a great number of places in various parts of the world, a more or less fluid inflammable matter exudes. It is known as Persian naphtha, Petroleum, Rock oil, Rangoon tar, Burmese naphtha, etc. These naphthas have been examined by many chemists, but the experiments have been exceedingly defective, and even the analyses most incorrect, for in all cases where a loss of carbon or hydrogen has been experienced, it has been put down as oxygen. The oil procured from the above sources, when rectified and well dried, contains no oxygen. The constitution of all of them is probably nearly the same, the odour and physical characters closely agreeing in specimens obtained from widely different sources. A thorough investigation of the most plentiful and well marked of all of these naphthas (namely, that from Rangoon) has been undertaken by MM. Warren De la Rue and Hugo Müller, who have been engaged upon it for some years. They find the fluid to consist of two principal series of hydrocarbons, namely, the benzole class and another, unacted upon by acids, and apparently consisting of the radicals of the alcohols. In addition to the fluid hydrocarbons, Burmese naphtha contains a considerable quantity of paraffine.

Burmese naphtha or Rangoon tar is obtained by sinking wells about 60 feet deep in the soil, the fluid gradually oozes in from the soil and is removed as soon as the quantity accumulated is sufficient. The crude substance is soft about the consistence of goose grease, with a greenish brown colour and a peculiar but by no means disagreeable odour. It contains only 4 per cent. of fixed matters. In the distillations, MM. De la Rue and Müller employed superheated steam for the higher, and ordinary steam for the lower temperatures. At a temperature of 212°, eleven per cent. of fluid hydrocarbons distil over; they are entirely free from paraffine. Between 230 degrees and 293° F. ten per cent. more fluid distils, containing, however, a very small quantity of paraffine. Between the last named temperature and 320° F., the distillate is very small in quantity, but from that to the fusing point of lead, 20 per cent, more is obtained. The latter, although containing an appreciable amount of paraffine, remains fluid at 32° F. At this epoch of the distillation, the products begin to solidify on cooling, and 31 per cent. of substance is obtained of sufficient consistency to be

---

[1] A. Laurent, Ann. de Chim. Phys. vol. II, 1837, pp. 66 & 139.

submitted to pressure. On raising the heat considerably, 21 per cent. of fluids and paraffine distil over. In the last stage of the operation, 3 per cent. of pitch-like matters are obtained. The residue in the still, consisting of coke containing a little earthy matter, amounts to 4 per cent. We thus have as the products in this very carefully conducted and instructive distillation.

Below 212°	Free from paraffine	11.0
230 to 293°	A little paraffine	10.0
293 to 320°		
320 to fusing point of lead	Containing paraffine, but still fluid at 320°	20.0
At about the fusing point of lead	Sufficiently solid to be submitted to pressure	31.0
Beyond fusing point of lead	Quantity of paraffine diminishes	21.0
Last distilled	Pitchy matters	3.0
Residue in still	Coke containing a little earthy impurity	4.0
		100.0

All the above distillates are lighter than water. Almost all the paraffine may be extracted from the distillates by exposing them to a freezing mixture. In this manner, no less than between 10 and 11 per cent. of this valuable solid hydrocarbon may be obtained from Burmese naphtha. We may before long expect a full account of the substances contained in Rangoon tar.

Naphtha appearing closely to resemble the above is found at Alfreton, Amiano (Duchy of Parma), Baku (borders of the Caspian), Barbadoes, Clermont (France), Gabian, near Beziers (France), Galicia, Neufchâtel (Switzerland), Tegernsee (Bavaria), Trinidad, United States, Val di Noto in Sicily, Wallachia, Zante, Mt. Zibio (Modena). Naphtha was one of the ingredients said, by some old authors, to enter into the composition of the Greek fire.

In 1857, our imports of Naphtha were 6,558 gallons; in 1858, 3,804 gallons; and in 1857, our exports were 164 gallons; and in 1858, 189 gallons.

NAPHTHA, Shale. The true constitution of shale naphtha, or, as it is sometimes called in commerce, "shale oil", has not yet been satisfactorily ascertained. In fact, to do so would involve a very laborious research, or rather series of researches, for the various shales or schists differ much in the quantities and qualities of the naphtha yielded by them. The bituminous shale of Dorsetshire contains much nitrogen and sulphur, arising to a great extent from presence of a large quantity of semi-fossilised animal remains. The crude naphtha, consequently, is intolerably foetid. By repeated treatments with concentrated sulphuric acid and caustic soda it may, however, be rendered very sweet. It then contains pretty nearly the same constituents as Boghead naphtha, *i.e.* benzole and its homologues, various hydrocarbons of the olefiant gas series, and small quantities of the alcohol radicals or isomeric hydrocarbons. There are also present, previous to purification, carbolic acid and numerous alkaloids; but, strange to say, in the samples I examined there were no traces of aniline to be found. There is little doubt that shales of this kind might be most profitably worked by one or other of the recently patented processes for the preparation of photogen and lubricating oil. Intimately connected with the oils of shale are the fluids yielded by the distillation of the numerous bitumens and asphalts found in various parts of the world. Undoubtedly these deposits will one day become of important use in the arts.

The bitumen of Trinidad yields on distillation an intensely foetid oil, and also a very large quantity of water. It also appears to give a considerable quantity of alkaloids and ammonia. It will, perhaps, scarcely be a profitable speculation at present to bring this bitumen so far for the purpose of distillation, but doubtless there are many ports into which it could be carried at a reasonable price. It is said that some has already found its way into America, for the purpose of having photogen prepared from it.

France is particularly rich in deposits of bitumen, especially in the volcanic districts of Auvergne. Switzerland, Italy, Germany, Russia, Poland, in fact almost every part of Europe contains bitumen of various degrees of consistency and value. Even in our own country there are deposits at Alfreton and other places. The Alfreton bitumen is not unlike that of Rangoon. Bitumens have been examined by various chemists, more especially by Boussingault and Voelckel. Their results, however, require to be repeated with great care, as hitherto sufficient attention has not been paid to the purification by chemical means of the various hydrocarbons. Fractional distillation, although absolutely necessary, in order to enable bodies to be obtained of different but specific boiling points, does not do away with the necessity for elaborate purifications by means of bromine, nitric, and sulphuric acids, etc. There is little doubt that a rigorous examination of the oils procurable by distillation of the various European and other bitumens would be rewarded, not only by scientific results of great interest, but also by discoveries of immense commercial importance. It must not be forgotten, in connection with the money value of such researches, that the bitumens yield a very high percentage of distillate, much greater than any of the shales or imperfectly fossilised coals which are wrought on the large scale for the preparation of illuminating or lubricating oils."

It will be seen from these extracts that the recent oil boom in Pennsylvania had not yet had its repercussions in European scientific and technical circles. The naphtha prepared from wood-tar is separately discussed in Ure's Dictionary, most of its constituents had been identified, but it did not play an important part in the speculations of Williams for the quantities of coal-tar and bitumens were already significantly larger.

Soon after Ure had printed his dictionary petroleum technology was to profit from the new theories of the organic chemists. No trace of such theories is yet found in the report which Professor Benjamin Silliman Jr., then Professor of General and Applied Chemistry at Yale, wrote on the sample of crude from a seepage on Hibbard farm near Titusville (in Venango County, Pennsylvania) sent to him by George H. Bissell and Jonathan G. Eveleth [1]. In this report Silliman first described the general properties of the crude oil and those of the fractions he had prepared by distillation and collected between

---

[1] Benjamin Silliman Jr., Report on the Rock Oil or Petroleum from Venango Co., Pennsylvania, with special reference to its use for illumination and other purposes (New Haven, 1855).

certain temperatures. He investigated the "liquidity" at low temperatures before and after refining, prepared illuminating gas by cracking the fractions and separated paraffin wax, which yielded excellent candles. He measured the candle power of the lamp oil, stressed the value of the lubricating oil fractions and ended his report with the remarkable words: "In conclusion, gentlemen, it appears to me that there is much encouragement in the belief that your company (The Pennsylvania Rock Oil Company of New York, founded on December 30, 1854) have in their possession a raw material from which, by a simple and not expensive process, they may manufacture very valuable products. It is worthy of note that my experiments prove that nearly the whole of the raw product may be manufactured without waste, and this solely by a well-directed process which is in practice one of the most simple of all chemical processes."

In 1863, however, Schorlemmer started a series of investigations into the constitution of petroleum [1] which continued over ten years. A year later Pelouze and Cahours examined American crudes which they identified as mixtures of paraffins [2], followed by another essay on Russian crudes in the same volume, samples of which they had obtained through the Admiral Likhatchoff from Baku.

Even more important were the investigations of Prof. H. E. Saint-Claire Deville (1818-1881) who directed the chemical laboratory at the International Exhibition of 1867 in Paris. As it happened several French inventors were at work devising the proper apparatus to burn the heavy residues of coal-tar and petroleum under steam engines and boilers. The Committee of the Exhibition would not allow them to demonstrate their inventions at the Exhibition itself but Saint-Claire Deville allowed Paul Audoin to show his invention in the laboratory he directed. The emperor Napoleon III was struck by these experiments and ordered Deville to study the possible applications of mineral oils as fuels at his expense after a thorough search on this subject had been carried out in literature. It happened that Napoleon III owned a yacht named "Puebla", which was fitted out with a 60 HP engine and boilers fired with coal, the soot of which annoyed his guests and spoiled the clothes of the ladies on board this ship.

Within a very short time Deville was able to have the survey of literature, carried out by his assistant C. Cogniet, published [3] and in the meantime he

---

[1] C. Schorlemmer, Die chemische Konstitution des Steinöls (Ann. d. Chemie vol. 127, 1863, pp. 311 ff.).

[2] J. Pelouze et A. Cahours, Sur les pétroles d'Amérique (Ann. Chim. Phys. vol. (4) I, 1864, pp. 5 ff.).

[3] C. Cogniet et H. Saint-Claire Deville, Des huiles minérales au point de vue de leur emploi pour le chauffage des machines à vapeur (Paris 1868).

collected samples from different countries. In the course of his investigations, in which he was assisted by his pupils Alfred Ditte (1843-1908), Maximilien Pougnet (1845-1880) and René-Valentin Prudhon (1841-1869), he analysed five crudes from Alsace, one from Gabian, three from Hanover, four from Italy, two from Galicia, two from Ploesti (Roumania), one from Greece, four from Canada, eight from the U.S.A., two from the Caucasus, one from Burma, three from Java and one from China. In each case he gave the elementary analysis, the lower part of the distillation curve, the specific gravities at 0°C and 50°C and the calorific value. For the latter value he used the methods described earlier by Favre and Silbermann [1].

Saint-Claire Deville soon published his results [2] in several essays which not only contained the figures obtained in his laboratory but also the results obtained with mineral oils as fuel for locomotives of the Chemins de Fer de l'Est [3] and on board the "Puebla" [4]. These were carried out with the help of his pupil Dieudonné (1836-1891) and Dupuy de Lôme using a special grate for burning oil designed by Deville himself.

These investigations did of course attract much attention in other European countries. In the Netherlands Prof. E. H. van Baumhauer, Professor of Chemistry at the University of Amsterdam was much impressed by these results. Acting on the work published by Pelouze and Cahours he had applied to the Minister of Colonies on May 6, 1863 to obtain samples of crudes from the seepages in the Indonesian archipelago. Only after he had analysed six samples did he read about Saint-Claire Deville's work, and he completed his analyses by determining the calorific values of his samples and certain other characteristics. His report [5] in which he urged the industrial exploitation of these petroleum resources was an important factor in the foundation of the Dordtsche Petroleum Maatschappij in Java and the Royal Dutch in Northern Sumatra a few decades later.

---

[1] Favre and Silbermann, Comptes Rendus vol. 23, 1846, pp. 200 ff.; Ann. Chim. Phys. vol. (3) 34, 1852, pp. 357 ff.

[2] H. Saint-Claire Deville, Sur les propriétés physiques et le pouvoir calorifique de quelques pétroles de l'empire russe (Comptes Rendus févr. 27, 1871, pp. 191-198).

H. Saint-Claire Deville, Premier mémoire sur les propriétés physiques et le pouvoir calorifique des pétroles et des huiles minérales (Comptes Rendus mars 9, 1868, pp. 442-454; mars 1, 1869, pp. 485-502).

[3] H. Saint-Claire Deville, De l'emploi industriel des huiles minérales pour le chauffage des machines et en particulier des machines locomotives (Comptes Rendus novembre 2, 1869, pp. 933-938).

A. Henry, La Sézanne (Chemins de Fer, Paris, mai/juin 1947, pp. 63 ff.).

[4] H. Saint-Claire Deville, Deuxième mémoire sur les propriétés physiques et le pouvoir calorifique de quelques pétroles (Comptes Rendus, févr. 15, 1869, pp. 349-357).

[5] E. H. von Baumhauer, Over de aardoliën der Nederlandsch Oost-Indische bezittingen (Verslagen Kon. Akad. Wet. Amsterdam, Afd. Natuurkunde vol. (2) III, 1876, pp. 340—383).

In 1872 Schorlemmer read a paper to the Chemical Society in which he explained the new chemistry of hydrocarbons based on the "tetrad carbon" atom which the majority of physicists and chemists now adopted. In the course of this lecture [1] he remarked:

"An investigation of the paraffins contained in Pennsylvanian petroleum showed that they belong to the first group, as the alcohols derived from them, as well as the acids obtained by oxidising the latter, were found to be normal compounds.
Before I take leave of the paraffins, I have to say a few words about the paraffins par excellence, *viz*. the solid paraffins. These bodies, which resist so energetically the action of chemical agents that they have obtained their name from this fact, appear to be very unstable bodies at a high temperature, although, curiously enough, they are produced by destructive distillation. Thorpe and Young have shown (Proceedings of the Royal Society, xix, 370), that by distilling solid paraffin under pressure, it is almost completely resolved, with evolution of but little gas, into hydrocarbons which remain liquid at the ordinary temperature, and consist principally of olefines."

An extensive article by L. Troot for Wurtz' chemical dictionary (1875) [2] embodied Saint-Claire Deville's results and gave details on the saturated hydrocarbons or paraffins present in petroleum. The part of the aromatics was recognized by that time and Beilstein and Kurbatow drew attention to the part of the cyclic hydrocarbons or "naphthenes" [3] a few years later. Hence the excellent handbooks by Aschan, Wischin, Ragousine and others published early in the twentieth century were already able to discuss the part which the different series of hydrocarbons played in the total characteristics of various crudes and methods for their separation and determination were already in use, in an imperfect form, by 1900.

As long as the chemists confined their efforts to the analysis of light petroleum fractions they could already expect some success with the tools at their disposal by 1880. Thus Allen read a paper to the British Association in 1879 on the composition of petroleum spirit and benzol giving the following table:

Petroleum Spirit, "Benzoline" or "Benzine"	Coal-Tar Naphtha, or "Benzol"
1. Consists of heptane, C_7H_{16}, and its homologues.	1. Consists of benzene, C_6H_6, and its homologues.

[1] C. Schorlemmer, The Chemistry of Hydrocarbons (J. Chem. Soc. vol. X, 1872, pp. 425-446).
[2] Article on "Pétrole", Ad. Wurtz, Dictionnaire de chimie pure et appliquée (Paris, 1873, part II, pp. 782-791).
[3] F. Beilstein & A. Kurbatow, Über die Natur des kaukasischen Petroleums (Berichte vol. 13, 1880, pp. 1818-1821); Über die Kohlenwasserstoffe des amerikanischen Petroleums (Berichte, 1880, pp. 2028-2029).

Petroleum Spirit,
"Benzoline" or "Benzine"

2. Heptane contains 84.0 per cent of carbon.
3. Commences to boil at 54° to 60°C.
4. Specific gravity at 15.5°C about 0.69 to 0.72.
5. Smells of petroleum.
6. Dissolves iodine, forming a solution of a raspberry-red colour.
7. Does not sensibly dissolve coal-tar pitch, and is scarcely coloured by it, even on prolonged contact.
8. When shaken in the cold, with one-third of its volume of fused crystals of absolute carbolic acid, the latter remains undissolved, and forms a separate lower stratum.
9. Requires two volumes of absolute alcohol, or four or five volumes of methylated spirit of 0.828 sp.gr., for complete solution at the ordinary temperature.
10. Warmed with four measures of nitric acid of 1.45 sp.gr. the acid is coloured brown, but the spirit is little acted on, and forms an upper layer.

Coal-Tar Naphtha, or
"Benzol"

2. Benzene contains 92.3 per cent of carbon.
3. Commences to boil at about 80°C.
4. Specific gravity about 0.88.
5. Smells of coal-tar.
6. Dissolves iodine, forming a purple-red liquid of the tint of an aqueous solution of potassium permanganate.
7. Readily dissolves coal-tar pitch, forming a deep-brown solution.
8. Miscible with absolute carbolic acid in all proportions.
9. Miscible with absolute alcohol in all proportions. Forms a homogeneous liquid with an equal measure of methylated spirit of 0.828 sp.gr.
10. Completely miscible with four measures of nitric acid of 1.45 sp.gr., with great rise of temperature and production of dark brown colour. A portion of the nitrobenzene may separate as the liquid cools.

and concluding that coal-tar naphtha consisted mainly of "benzol or benzene" whereas in petroleum spirit the "proportion of heptane, present in commercial benzoline, probably equalled, or even exceeded, that of all the other constituents" [1].

In 1880 Allen read a second paper to the British Association discussing the chemical composition of the different fractions of petroleum and shale naphthas [2] and giving the following summary:

"Shale naphtha contains a large proportion of olefines or hydrocarbons of the general formula C_nH_{2n}, while petroleum spirit consists chiefly of paraffins or

---

[1] Alfred H. Allen, Notes on Petroleum Spirit or "benzoline" (Chemical News vol. XL, 1879, pp. 101-102).
[2] Alfred H. Allen, Further Notes on Petroleum Spirit and Analogous Liquids (Trans. Brit. Assoc. Advancement of Science 1880, pp. 547-548).

hydrocarbons of the formula C_nH_{2n+2}. Quantitative experiments with nitric acid show that while shale naphtha contains only 15 to 25 per cent. of paraffins, the balance being olefines, in petroleum spirit these proportions are reversed.

The burning oil from shale also consists chiefly of olefines, while in kerosene or petroleum burning oil paraffins predominate.

The following table shows the general composition of the products from shale and petroleum:

	Petroleum	Shale
Naphtha	At least 80 per cent. paraffins, the remainder probably olefines. Distinct traces of benzene and its homologues.	75 to 90 per cent. of paraffins, the remainder olefines. No benzene or its homologues.
Photogene	55 to 80 per cent. of higher paraffins, the remainder apparently olefines.	Apparently 60 to 65 per cent. of olefines, the rest paraffins.
Lubricating oil		Consists almost wholly of higher olefines. No naphthalene.
Wax	Solid paraffins.	Solid paraffins."

Saint-Claire Deville, Mendelejeff and other contemporaries had already stressed the necessity of identifying properly organic compounds such as the constituents of petroleum. One of the best methods was the determination of their physical characteristics such as specific gravity, boiling point, etc. Indeed much work was done in this direction during the last three decades of the nineteenth century, but the refiners did not always follow suit. This can be illustrated by the unfortunate adoption of the Baumé scale for expressing the specific gravity of petroleum products and the confusion which followed, which could have been avoided by using the normal hydrometric scale already adopted by all organic chemists about 1860.

Baumé had devised an hydrometer scale divided in degrees for expressing the specific gravity which he first published in "L'Avant-Coureur" of November 7, 1768 and then in his Eléments de Pharmacie of 1769. This scale was meant for liquids heavier than water. Louis-Benjamin Francoeur (1773-1849) devised a similar scale for liquids lighter than water [1]. The modulus he used in his formula was 146, but the American chemist H. Pemberton, who introduced this formula in his own country adopted a modulus 140, which entered in the U. S. Dispensatory. The Baumé scale was adopted by the young American petroleum industry, but the American Petroleum Association in 1864 re-introduced the modulus 146 in the formula and thus was in conflict

---

[1] L. B. Francoeur, Mémoire sur l'aréométrie (Paris, 1842).

with the formula of the U. S. Dispensatory to which a number of refiners and oil merchants clung. These different views attracted the attention of the American Chemical Society in 1876 and Prof. Chandler of Columbia University investigated the matter and agreed that the modulus 146 was correct [1].

About 1900 two different scales, one with a modulus of 140 and one with 141.5 still continued in use [2] and others used a Tagliabue scale to express the specific gravity of petroleum. At the First World Petroleum Congress of Paris (1900) Boverton Redwood rightly drew the attention to this distressing situation and he proposed unification. Unfortunately the U.S. Bureau of Standards did not succeed in establishing unification before 1922, though the Germans had succeeded in standardizing the test scale of American crudes in 1892 [3] and had published tables for Russian crudes in 1893.

Thus the petroleum industry had still a long way to go after the years in which the chemists had established the constituents of crude oils. The struggle for proper methods to determine the chemical and physical properties of petroleum products and for internationally recognized units in which such properties were expressed is an epic in itself.

---

[1] C. F. Chandler, The Baumé hydrometer (Mem. Acad. Sci. vol. III, i, 1884, pp. 63-71).
[2] U. S. Standard Baumé Scales (Circular No. C. 59; U. S. Bureau of Standards, April 5, 1916).
[3] Mitt. K. Normal-Aichungs Kommission Berlin, No. 17, Januar 27, 1892.

CHAPTER FOUR

NAPHTHA GOES TO WAR

When, two centuries ago the first Russian governor of the Caucasian region sent a sample of crude oil to the Academy of St. Petersburg, the learned members wrote him that "this evil-smelling stuff can not be used for anything but the lubrication of wheels". Unfortunately they had not consulted their classics, for they would soon have learnt from these ancient documents that petroleum had not only been put to various uses, but that it had even been employed in warfare.

The use of fire in warfare is probably as old as war itself. Assyrian reliefs show pictures of warriors pouring boiling oil from city walls, using torches or extinguishing fires kindled by the enemy. Homer tells us [1], that "the Trojans cast upon the swift ship unwearied fire, and over her forthwith streamed a flame that might no be quenched", but this is not such a strange fire after all, for Patroclus succeeds in putting out this fire after some time. Again after the battle of Salamis the Greeks "followed after the Persians in their flight, hewing them down, till they came to the sea. There they called for fire and laid hands on the ships" [2].

Soon inflammable substances were added to the faggots and fire brands. Thus the Peloponnesians besieging Plataea in 429 B.C. made a final attempt to take this city [3]:

"After this, the Peloponnesians, seeing that their engines were doing no good and that the counter-wall was keeping pace with the mound, and concluding that it was impracticable without more formidable means of attack to take the city began to make preparations to throw a wall about it. But before doing so they decided to try fire, in the hope, that if a wind should spring up, they might be able to set the city on fire, as it was not large; indeed, there was no expedient they did not consider, that they might if possible reduce the city without the expense of a siege. Accordingly they brought faggots of brushwood and threw them down from the mound, first into the space between the wall and the mound; and then, since the space was soon filled up by the multitude of workers, they heaped faggots also as far into the

[1] Homer, Iliad XVI. 123-124.
[2] Herodotus VI. 113.
[3] Thucydides II. 77.

city as they could reach from the height, and finally threw fire together with sulphur and pitch upon the wood and set it afire. And a conflagration arose greater than any one had ever seen up to that time, kindled, I mean, by the hand of man; for in times past in the mountains when dry branches have been rubbed against each other a forest caught fire spontaneously therefrom and produced a conflagration. And this fire was not only a great one, but also very nearly destroyed the Plataeans after they had escaped all earlier perils; for in a large part of the city it was not possible to get near the fire, and if on top of that a breeze had sprung up blowing toward the city, which was precisely what the enemy were hoping for, the Plataeans would not have escaped. But as it was, this also is said to have happened—a heavy thunder-shower came on and quenched the flames, and so the danger was checked".

Fire in its simple form of torches continues to form an element in warfare. At Fidenae (426 B.C.) "Fire was the weapon of that vast multitude, and blazing torches threw a glare upon the entire throng" [1], and the works of historians like Thucydides, Diodor, Appian and others abound with passages of this type. At sea fireships appear fairly early. During the siege of Syracuse [2] "the Syracusans, wishing to set the Athenian ships afire, turned loose an old merchant ship which they filled with faggots and pine-wood, after casting fire into it, the wind being in the direction of the Athenians. And the Athenians, alarmed for their ships, devised in their turn means for hindering and quenching the flames and having stopped the fire and prevented the ship from coming near, escaped the danger". For centuries to come this was a regular feature of sea-warfare. Procopius many centuries later [3] tells us that "when they (the Romans) came near, the Vandals set fire to the boats which they were towing, when their sails were bellied by the wind, and let them go against the Roman fleet (of Basilicus)".

When Cyrus plans to enter the town of Babylon and his generals warn him of possible street fighting he answers: "If any do go upon their houses, we have a god on our side, Hephaestus. And their porticoes are very inflammable, for the doors are made of palm-wood and covered with bituminous varnish which will burn like tinder; while we, on our side, have plenty of pine-wood for torches, which will quickly produce a mighty conflagration; we also have plenty of pitch and tow, which will quickly spread the flames everywhere, so that those upon the house-tops must either quickly leave their posts or quickly be consumed" [4]. When the Roman consul M. Fulvius besieged Ambracia the Aetolians counter-attacked, "some advanced with flaming torches, others carrying tow and pitch and fire-brands, the whole battle-line gleaming with

---

[1] Livy IV. xxxiii. 2.
[2] Thucydides VII. liii. 4.
[3] Procopius III. vi. 18-22 (Vandalic Wars).
[4] Xenophon, Cyropaedia VII. v. 23.

flames"[1]. Some defence was possible against such artificial fires for, defending Byzantium in A.D. 194,

"Severus, learning of Priscus' skill, prevented his execution, and later made use of his services on various occasions, especially at the siege of Hatra, where his machines were the only ones not burned by the barbarians"[2].

Still the use of sulphur and tow and similar devices continued for when

"Maximinus began to invest Aquilea closely ..... the townsmen defended themselves from the soldiers with sulphur, fire and other defensive devices of the same kind; and of the soldiers some were stripped of their arms, others had their clothing burned, and some were blinded, while the investing engines were completely destroyed"[3].

Hence military experts like Aeneas the Tactician write instructions such as [4]:

"XLV. One must pour pitch and cast sulphur, then set on fire a faggot and let it down by a rope upon the particular object we wish. And such things as these, held out from the places in which we are standing, are hurled at the approaching engines",

or instructing his readers to use vinegar to prevent such fires:

"..... And if you know in advance the parts that are likely to be set on fire, rub vinegar on the outside and the flame will not advance on them".

Of course means were soon found to set more distant goals on fire. One of the earliest forms of such incendiary weapons were fire-arrows, which the Greeks had to fight when the Persians attacked their home country [5].

"The Persians encamped upon the hill over against the citadel, which is called Areopagus by the Athenians, and began the siege of the place, attacking the Greeks with arrows whereto pieces of lighted tow were attached, which they shot at the barricade".

Such fire-arrows seem to have been invented in the East. Not only does Philostratus tell us [6] that:

"The inhabitants of the Caucasus regard the eagle as a hostile bird, and burn out the nests which they build among the rocks by hurling into them fiery darts",

---

[1] Livy, Roman History XXXVIII. v. 2.
[2] Dio Cassius LXXV. 11. 2.
[3] Historia Augusta: Capitolinus, the Two Maximini xxii. 5-6.
[4] Aeneas Tacticus XXXVIII and XLV.
[5] Herodotus VIII. 52.
[6] Philostratus, Life of Apollonius of Tyana II. cap. iii.

but we have many stories of their use by the Indian King Porus and his army when Alexander the Great tried to penetrate into that country. Such fire-arrows soon appear in many regions of the Mediterranean. The Romans find them in use in Spain [1]:

"The Saguntines had a javelin, called a falarica, with a shaft of fir, which was round except at the end whence the iron projected; this part, four-sided as in the pilum they wrapped with tow and smeared with pitch. Now the iron was three feet long, that it might be able to go through both shield and body. But what chiefly made it terrible, even if it stuck fast in the shield and did not penetrate the body, was this, that when it had been lighted at the middle and so hurled, the flames were fanned to a fiercer heat by its very motion, and it forced the soldier to let go his shield, and left him unprotected against the blows that followed",

and incorporate this weapon in their own army as this passage from Plutarch will show [2]:

"Sulla, seeing what was going on (in Rome), shouted orders to set fire to the houses and seizing a blazing torch, led the way himself, and ordered his archers to use their fire-bolts and shoot them up at the roofs".

Demetrius in his siege of Rhodus (305 B.C.) used such fire-arrows in an attempt to burn the Rhodian ships but without success [3].

About 350 A.D. such fire-arrows are common subjects for tacticians and the army handbooks of the Romans describe several types as the following passages from Ammianus [4] and Aeneas Tacticus [5] will show:

"Fire-darts (malleoli) are made in this form: the shaft is of reed, and between this and the point is a covering of bands of iron; it looks like a woman's distaff for making linen threads. It is skilfully hollowed out on the lower side with many openings, and in the cavity fire and some inflammable matter are placed. And if it is shot slowly from a somewhat soft bow (for it is extinguished by too swift a flight) and has stuck anywhere, it burns persistently, and water poured upon it is no way of extinguishing it except by sprinkling it with dust, for water rouses the fire to still greater heat." [6]

and

"You must pour pitch and cast tow and sulphur on the pent-houses that have been brought up, and then a faggot fastened to a cord must be let down in flames upon the pent-house. And such things as these, held out from the walls, are hurled at the engines as they are being moved up, by which the latter are to be set on fire. Let sticks be prepared shaped like pestles but much larger, and into the ends of each

---

[1] Livy XXI. viii. 9.
[2] Plutarch, Sulla IX. 7.
[3] Diodorus Siculus XX. 88.2.
[4] Ammianus Marcellinus XXIII. 4. 14-15.
[5] Aeneas The Tactician XXXIII-XXXV.
[6] See note 4.

stick drive sharp irons, larger and smaller, and around the other parts of the stick, above and below, separately, place powerful combustibles. In appearance it should be like bolts of lightening as drawn by artists. Let this be dropped upon the engine as it is being pushed up, fashioned so as to stick into it, and so that the fire will last after the stick has been made fast. Then, if there are any wooden towers, or if part of the wall is of wood, covers of felt or raw hide must be provided to protect the parapet so that they cannot be ignited by the enemy. If the gate is set on fire you must bring up wood and throw it on to make as large a fire as possible, until a trench can be dug inside and a counter-defence be quickly built from the materials you have at hand, and if you have none, then by tearing down the nearest houses.

If the enemy tries to set anything on fire with a powerful incendiary equipment you must put out the fire with vinegar, for then it cannot easily be ignited again, or rather it should be smeared beforehand with birdlime, for this does not catch fire. Those who put out the fire from places above it must have a protection for the face, so that they will be less annoyed when the flame darts towards them.

And fire itself which is to be powerful and quite inextinguishable is to be prepared as follows. Pitch, sulphur, tow, granulated frankincense, and pine sawdust in sacks you should ignite and bring up if you wish to set any of the enemy's works on fire."

As new means were found primitive forms of flame throwers were developed such as were used during the siege of Delium in 425 B.C. [1]:

"The Boeotians .... made an expedition against Delium and attacked the fortification. After trying other forms of assault they took it by bringing up an engine made in the following manner. Having sawed in two a great beam they hollowed it throughout, and fitted it together nicely like a pipe; then they hung a cauldron at one end of it with chains, and into the cauldron an iron bellowspipe was bent from the beam, which was itself in a great part plated with iron. This engine they brought

Fig. 2. The Flame-thrower described by Thucydides (Sixth century B.C.).

---

[1] Thucydides IV. 100.

up from a distance on carts to the part of the wall where it was built chiefly of vines and wood; and when it was near they inserted a large bellows into the end of the beam next to them and blew through it. And the blast passing through the air-tight tube into the cauldron, which contained lighted coals, sulphur, and pitch, made a great blaze and set fire to the wall, so that no one could stay on it any longer, but all left it and took to flight; and in this way the fortification was taken".

and we hear of similar instruments used in naval warfare about 190 B.C. [1]:

"The engine for throwing fire (pyrphoros) used by Pausistratus, the Rhodian admiral, was funnel-shaped. On each side of the ship's prow noosed ropes were run along the inner side of the hull, into which were fitted poles stretching out seawards. From the extremity of each hung by an iron chain the funnel-shaped vessel full of fire, so that, in charging or passing, the fire was shot out of it into the enemy's ship, but was a long way from one's own ship owing to the inclination".

These and other passages on "chemical fires" were examined by military experts [2] whose writings should, however, be used with great care as the references and translations are not always up to standard. Some of them have tried to argue that ancient stories about fires called down from heaven by the prophets or magicians might have been such chemical fires. Of course the Old Testament is full of passages referring to "the fire that came out from before the Lord" but we have hardly any reason to connect these with the "fires" of the tacticians.

Such passages belong to an entirely different world, parallels can be found in the stories which circulated in the ancient world about the power of the Brahmins, "the genuine sages who live between the Hyphasis and the Ganges", of whom it is said that [3]:

"they do not battle with those who approach them, but they repulse them with prodigies and thunderbolts which they send forth, for they are holy men and beloved of the gods. It is related, anyhow, that Hercules of Egypt and Dionysus after they had overrun the Indian people with their arms, at last attacked them in company, and that they constructed engines of war, and tried to take the place by assault; but the sages, instead of taking the field against them lay passive and quiet, as it seemed to the enemy; but as soon as the latter approached they were driven off by rockets of fire and thunderbolts which were hurled obliquely from above and fell upon their armour".

---

[1] Polybius, Histories XXI. 7.
[2] H. Hecht, Das Erdöl als Kriegsmittel bis zur Erfindung des Schiesspulvers (Erdöl und Kohle 1943, No. 5, pp. 117-128).
R. Pique, Histoire de la pyrotechnie de guerre (Chimie et Industrie, Compte Rendu du IXe Congrès de Chimie Industrielle, pp. 347-358).
A. Hausenstein, Von den Vorläufern des Flammenwerfers (Z. gesammt. Schiess- u. Sprengstoffwesens vol. XXXIV, 1939, pp. 10-14, 44-47; 73-77; 105-109).
[3] Philostratus, The Life of Apollonius of Tyana II. 33

We know that certain machines existed in Greek and Roman theatres to imitate thunder and lightning, like the one which Caligula had made [1] when he wanted to show that he was mightier than Jupiter:

"Caligula had a contrivance by which he gave answering peals when it thundered and sent return flashes when it lightened".

Such stories of "artificial fires" being used in the Old Testament begin with the claim of Roger Bacon in his De mirabili potestate artis et naturae (1265) that Gideon in his battle against the Midianites used bottles containing a certain mysterious fire, the neck of which was broken on the battlefield and lit before they were thrown at the enemy. Such stories cannot be of any help to reconstruct the history of chemical warfare.

It is more interesting to study the composition of the incendiary mixtures which developed beyond the sulphur-pitch-tow formula. Resins, juices of certain plants and roots or animal products were added from time to time. Pliny draws the attention to such a product [2]:

"The same Pythagoras calls aproxis, a plant whose root catches fire at a distance like naphtha".

Some of these ingredients were said to be self-igniting, and therefore eminently suitable for such compositions. Thus Philostratus tells us [3]:

"There is also a creature in this river (Hyphasis in India) which resembles a white worm. By melting this down they make an oil, and from this oil, it appears, there is given off a flame such that nothing but glass can contain it. And this creature may be caught by the king alone who utilises it for the capture of cities; for as soon as the fat in question touches the battlements, a fire is kindled which defies all the ordinary means devised by men against combustibles".

Such ingredients did in fact belong to the magician's toolbox [4]:

"Xenophon the juggler was also held in great admiration. He left behind him a pupil, Cratisthenes of Phlius, who could make fire burn spontaneously and invented many other magical tricks to confound man's understanding",

but they might be applied with profit to fire-arrows, incendiary mixtures and flame throwers. As the ancients were fully aware of the possibilities of light, inflammable crude oils and naphtha, we need not wonder that these are to be found in such recipes from the turn of our era onwards. We already have an example of their self-igniting properties in a passage of the Book of the Maccabees [5]:

[1] Dio Cassius, Roman Hstory LIX. 28. 6.
[2] Pliny, Natural History XXIV. ci. 158.
[3] Philostratus, Life of Apollonius of Tyana III. cap. 1.
[4] Athenaeus, Deipnosophistae I. 19e.
[5] II. Macc. 1. 19-22.

"For when our fathers were led into Persia, the priests that were then devout took the fire of the altar privily, and hid it in a hollow place of a pit without water, where they kept it sure, so that the place was unknown to all men. Now after many years, when it pleased God, Neemias being sent from the king of Persia, did send of posterity of those priests that had hid it to the fire: but then they told us they found no fire, but thick water. Then commanded he them to draw it up, and to bring it; and when the sacrifices were laid on, Neemias commanded the priests to sprinkle the wood and the things laid thereupon with the water. When this was done, and the time came that the sun shone, which afore was hid in the cloud, there was a great fire kindled, so that every man marvelled".

During the first century B. C. naphtha was used more frequently by the inhabitants of cities in Asia Minor, when they were attacked by the Romans. Here we have the testimonies of Pliny [1]:

"In Samosata the capital of Commagene (on the W. bank of the Euphrates) there is a marsh that produces an inflammable mud called petroleum (maltha). When it touches anything solid it sticks to it; also when people touch it, it actually follows them as they try to get away from it. By these means they defended the city walls when attacked by Lucullus (in the Mithridatic War, 74 B.C.): the troops kept getting burnt by their own weapons. Water merely makes it burn more fiercely; experiments have shown that it can only be put out by earth . . . . .
Naphtha is of a similar nature . . . . . it has a close affinity with fire, which leaps to it at once when it sees it in any direction."

and of Dio Cassius [2], who describes Lucullus' siege of Tigranocerta in 69 B.C.:

"But the barbarians did him serious injury by means of their archery as well as by the naphtha which they poured over his engines; this chemical (pharmakon) is full of bitumen and is so fiery that it is sure to burn up whatever it touches and it cannot easily be extinguished by any liquid".

Herodianus tells us that when Maximinus I besieged Aquilea "the defendants of the town threw down rocks and flung a mixture of sulphur, pitch and petroleum, in hollow vessels with long handles, after lighting it, towards the approaching, attacking army (probably with the help of ballistae) and covered it as it were with a thick, incessant shower of fiery rain". As the Romans did not possess the oil seepages of Mesopotamia and Persia the use of petroleum was rather restricted, for it was difficult to get this "oleum incendiarum" or "oleum Medicum" (Median (Persian) oil). Hence the military authors usually describe its use in the Oriental armies. Procopius tells us [3]:

"The Persians invented this: having filled jugs with sulphur, bitumen and the drug called naphtha by the Medes and Median oil by the Greeks, and having lighted them, flung them towards the rams and siege engines, and succeeded in burning

---

[1] Pliny, Natural History II. cviii.
[2] Dio Cassius, Roman History XXXVI.
[3] Procopius, Gothic Wars Lib. IV. cap. 11.

them in a short time . . . . . for this fire consumes the objects it touches, if it is not removed immediately"

and we also have the testimony of Ammianus [1], who has defined in an earlier passage (xxiii. 6. 16) naphtha to be a "glutinous substance which looks like pitch and which is similar to bitumen". He writes:

"In this neighbourhood (Media) the Medic oil (oleum Medicum) is made. If a missile is smeared with this oil and shot somewhat slowly from a loosened bow (for it is extinguished by a swift flight), wherever it lands it burns persistently; and if one tries to put it out with water, he makes it burn more fiercely, and it can be quelled in no other way than by throwing dust upon it. Now, the oil is made in this way. Those who are skilled in such matters take oil of general use, mix it with a certain herb and let it stand for a long time and thicken, until it gets magic power from the material. Another kind, like a thicker kind of oil, is native to Persia and (as I have said) is called in that language naphtha".

But some like Vegetius [2] instruct their readers to use it: "The fire-arrows set afire everything as they arrive burning. However, the falarica, has the form of a lance with a strong iron head in front, which is wrapped with tow, drenched in sulphur, resin, bitumen and "incendiary oil" (petroleum)". Such arrows have the form of a distaff and they are often shown in later medieval pictures of warriors.

When naphtha was not available cracked vegetable oils, called "empyreumatic oils" by the medieval authors, were sometimes used. Such oils, made by dry distillation (and cracking) of say olive oil, contained highly inflammable, volatile products and though they were mostly used for pharmaceutical and medical recipes they do occur in recipes for incendiary mixtures such as that from the manuscript of Conrad Kyeser of Eichstätt, written in 1405:

"OLEUM BENEDICTUM.

If you would make Oleum Benedictum take old olive oil and a whole red tile which has been washed in fresh water. Break half the stone into small pieces to the extent of a handful, make them glow in the fire so that they become as red as may be, then take a nice piece separately and suspend it in the oil and take and pound the piece and break it up—then (take) a selected jar which will stand the fire and fill it with tartar and the oil; next close the jar with Lutum Sapientiae and suspend it a while in the oven until the cement is dry. Then make a little fire and make it bigger and bigger so long as the water (*i.e.* the liquid) issues from it. Then stoke up the fire strongly until you see the red oil flowing. But take care that the oil does not get into the fire, for it cannot be put out. Go on stoking the fire until no more oil comes away, then allow the oven to cool and take care of it forthwith".

We can trace the gradual evolution of the ancient sulphur-pitch-tow mixture into modern types of gunpowder by the addition of several new ingredients

---

[1] Ammianus Marcellinus XXIII. 6. 37-38.
[2] Vegetius, De Re Militari IV. cap. 18.

such as petroleum and saltpetre. This is clear from a table given by Hime [1]:

Aeneas c. 350 B.C.	Vegetius c. 350 A.D.	Marcus Graecus Liber Ignium 1200-25	Kyeser 1405	Whitehorne 1560	Carcass composition 1903
Sulphur	Sulphur	Sulphur	Sulphur	Sulphur	Sulphur
Pitch	Bitumen	Pitch	Petroleum	Pitch	Tallow
Pinewood	Rosin	Sarcocolia	Salfanium(?)	Rosin	Rosin
Incense	Naphtha	Petroleum	Saltpetre	Turpentine	Turpentine
Tow		Sal coctus (refined saltpetre) Oil of Gemma Tartarum (bitartrate of Potash)		Bay Salt Saltpetre	Crude antimony Saltpetre

We need not wonder that saltpetre came to be used in such mixtures side by side with petroleum or petroleum fractions, for saltpetre had been known for many centuries, though we hear only of some limited application in ancient Assyria [2].

The main ingredient of such inflammable mixtures, petroleum, grew more and more important. A Byzantine anonym writing about 560 A.D. mentions "fire propelled with syringes (klysteros)" and he says that naphtha was as important to the Byzantines as iron and more important even than gold and silver. Hence the importance of the contacts with Caucasia, the region from which the Median oil came [3]. Some of these mixtures were of the nature of explosives. We are told that Leo I in his campaign against the Vandals "had little tubes thrown at the enemy filled with fire, which tubes often exploded in the hands of those who were to throw them". Agathias (in his life of Justinian) tells us that Anthemios of Tralles, builder of the St. Sophia church at Byzantium and a chemist by profession destroyed the house of his neighbour, the rethor Zenon, by "sending him thunder and lightening".

A new development took place in the seventh century. By 650 A.D. the Arabs had conquered the larger part of the Near East and were trying to become masters of the sea. Towards the end of the reign of the Caliph Mu'awiya an Arab fleet moved towards Constantinople, wintering at Cyzicus on the Asiatic coast of the Sea of Marmora and renewing its attack on the capital each spring for five years. This fleet was destroyed by the emperor Constantine

---

[1] H. W. L. Hime, Origin of Artillery (London, 1915, p. 28).
[2] R. J. Forbes, Studies in Ancient Technology, vol. III, p. 181 (Leiden, 1955).
[3] C. Toumanoff, Caucasia and Byzantine Studies (Traditio, vol. XII, 1956, pp. 409-425).

IV Pogonatus in 673 with the help of a new "fire" discovered a year earlier by an architect named Kallinikos. This is what the contemporary historians had to say on this point:

"Constantine, being apprised of the designs of the unbelievers against Constantinople, command large boats equipped with cauldrons of fire and fast-sailing galleys equipped with siphons.

At this time Kallinikos, an architect from Heliopolis of Syria, came to the Byzantines and having prepared a "marine fire" (pyr thalassion) set fire to the boats of the Arabs, and burned these with their men aboard, and in this manner the Byzantines were victorious and found the marine fire" [1].

"Then Kallinikos, an architect from Heliopolis of Egypt, came to the Byzantines and having made a "marine fire" burned and sank with every soul aboard the ships of the Arabs in Cyzicus. This is the man who discovered the "marine fire" and from him is descended the family of Lampros which today compounds the "artificial fire (pyr skeuaston)" [2],

or Constantine VII Porphyrogenitus (912-959) reporting to his son Romanus:

"Let it be known that during the reign of Constantine Pogonatus, one Kallinikos having fled from Heliopolis to the Byzantines, prepared the liquid fire, which by siphons (is ejected) and through which the Byzantines burned the fleet of the Arabs in Cyzicus (and) obtained victory" [3].

This new "Greek fire" served the Empire well. The recipe of Kallinikos, the Syrian (for most authors agree that he did not come from Egypt as Cedrenus claimed) was used by the emperor Leo III when he attacked the Arab fleet in 717 "while hunger and the extremely severe winter of the year 717-718 completed the final defeat of the Moslems", which saved not only the Eastern Empire of Byzantium, but also effectively protected the eastern flank of Western civilisation. It was the equivalent of the battle of Poitiers fought on the western front (732 A.D.).

Of the many naval battles in which the Byzantines used this Greek fire we will mention only the attack on Byzantium during the reign of Constantine Porphyrogenitus by the Russian fleet of some 1,000 vessels commanded by prince Igor which was defeated by a small squadron of 16 Byzantine ships provided with "syphons" on prow, stern and sides (941 A.D.). The Greeks profited by good weather, for Liutbrand [4] says: "God calmed the winds and quietened the sea for the circumstances were unfavourable for the Greeks

---

[1] Theophanes, Chronographia, A. M. 6164-6165 (Migne, Patrologiae Graeca, Tomus CVIII, pp. 720-721).

[2] Cedrenus, Synopsis Historion, ed. Becker, Bonn, 1839, p. 765.

[3] Constantine Porphyrogenitus, De administrando imperio, cap. 48, 20; Zonaros, Chronikon, vol. II, lib. xiv, p. 90.

[4] Liutbrand Book V. cap. 6.

to throw the fire". Few Russian ships escaped and the Russian Chronicle of Nestor had this excuse: "Then armed by a winged fire and by means of certain tubes the Greek general flung the flames onto the Russian warships, a terrible and marvelous spectacle! The Russians, perceiving this magic fire, fled towards the sea to escape its touch and a small number managed to regain the homeland. On their return they said to their compatriots: The Greeks have a fire which flies through the air like lightening; they threw it on us and burnt our vessels; hence we were not able to beat them".

We have but few data to go on for a discussion of the composition of this Greek fire and the way in which it was applied. Our most important source is the treatise on military tactics written by the emperor Leo VI (886-911) [1], which contains the following passages:

"8. ..... and of the last two oarsmen in the bow, let the one be the siphonator, and the other to cast the anchors into the sea.

6. In any case, let him have in the bow the siphon covered with copper, as usual, by means of which he shall shoot the prepared fire upon the enemy. And above such siphon (let there be) a false bottom of planks also surrounded by boards, in which the warriors shall stand to meet the oncoming foes.

45. On occasion (let there be) formations immediately to the front (without manoeuvers) so whenever there is need to fall on the enemy at the bow and set fire to the ships by means of the fire of the siphons.

51. Many very suitable contrivances were invented by the ancients and moderns, with regard to both the enemy's ships and the warriors on them—such as at that time the "prepared fire" which is ejected (thrown) by means of siphons with a roar and a lurid (burning) smoke and filling them (the ships) with smoke.

70. They shall use also the other method of small siphons thrown by hand from behind iron shields, and held (by the soldiers) which are called "hand siphons" and have been recently manufactured by our state. For these can also throw (shoot) the prepared fire into the faces of the enemy.

53. And against the scaling ladders which are brought up to the walls or against those who dare approach, (throw) stones and boiling pitch and oil and "tilis" .... Against the wooden towers, however, which are brought up on rollers and which are called by the tactician "mosynes", the so-called „strepta" is useful, which ejects the liquid fire (many indeed call this "brilliant") mechanically. And also the so-called handsiphons which our royal state has now developed (are useful)".

Besides we have some evidence from the chronicles of the Princess Anna Comnena, who describes a battle of the Byzantines against the Pisans (near Rhodes in 1103) and gives a recipe for such a "fire" in these words [2]:

"Captain Eleemon impudently attacking a very large ship at the stern, and falling on its rudders and, not being able readily to make his way through there, would have

---

[1] Leonis Imperatoris Tactica, edit. Meursius (Opera Meursii, Florence, 1745, pp. 529-920). Leo's Taktika (J. P. Migne, Patrologiae Graeca, Tomus CVII, chap. 19).
[2] Anna Comnena, Alexias, cap. XI & XIII.

been captured had he not promptly looked after his equipment and shot fire at them with a good aim. Then quickly veering about he immediately set on fire three very large ships of the barbarians ..... the barbarians were thoroughly frightened, for one thing on account of the "fire" hurled at them (for they were not accustomed to such implements or fire, naturally having an upward tendency, yet being shot by the shooter at whatever he might wish both downward and frequently to either side)."

"One collects pitch and the highly combustible juice of certain green trees. One grinds the mixture with sulphur and presses it into small reed tubes into which it is introduced by a strong and continuous wind like that of a flute-player. One lights them by applying fire to the exterior and like a burning meteor it falls on the objects in its way. The inhabitants of Durazzo (besieged by Bohemund in 1104) have used this fire when they were face to face with their enemies and burnt their beards and faces".

Of course such recipes were considered a most important state-secret. The emperor Constantine Porphyrogenitus impresses this on his readers in the following words [1]:

'You must of all things spend your care and your attention on the liquid fire which is launched with the help of tubes; and if they dare ask you for it, as they often did ask me, you must deny and reject this demand answering that this fire was shown and revealed by an angel to the great and holy first Christian emperor Constantine. With this message and by the angel himself he was bound, according to the authentic tradition of our ancestors, not to prepare this fire but for Christians, only in the imperial city and never anywhere else; also not to transmit it and never to reveal it to any other nation, whatever it might be. Then the great emperor, in order to protect himself against his successors, had graven on the holy table of God's church bans against anyone who would dare to reveal it to strangers. He prescribed that the traitor should be considered unworthy of the name Christian, of all posts and honours; whatever rank he held should be taken from him. He declared anathema for ever, he declared infamous whoever, emperor, patriarch, prince or subject, who would try to violate this law. He also ordered all men who feared and loved God to treat the malefactor as a public enemy, to condemn him and to deliver him to the cruelest torture. Still once it happened (crimes slip in everywhere) that one of the grandees of the Empire, won by immense presents, communicated this fire to a stranger, but God could not let such a crime go unpunished and one day as the culprit was about to enter the holy church of Saint Sophia a flame descended from heaven and devoured him. All present were seized with terror, and no one since has dared, whatever his rank be, to dream of such a crime or even less to execute it".

This "secret weapon" of the Byzantines, the "liquid fire" (pyr hygron) which according to Leo the Deacon [2] could "even incinerate the very stones", was certainly a compound containing naphtha, either in the form of crude oil or in that of distilled low-boiling fractions, such as the Arabs prepared for the

---

[1] Constantine Porphyrogenitus, De Administrando Imperio, Cap. 13.
[2] Leonis Diaconi Historia, edit. M. Hase, Leipzig, 1819, I, IX, c. x, p. 96c.

dry cleaning of silk and other valuable textiles. Though its secret leaked out slowly and we find Arabs using it too (or at least similar compounds) at an early date, we do not hear of its composition from early Latin manuscripts except in one case [1]. This late ninth-century Latin document contains the following passage:

"The composition of the fire of the three sages, naphtha, tow and tar (as used for fire arrows. Naphtha is a kind of balsam (*i.e.*, aromatic resin), having its source in Mount Sinai (and) exuding from a rock hence (called) oil rock. The other (balsam is extracted) from twigs. Combined, they make a fire which cannot be put out. For when the Saracens opened their campaign with a naval battle (their opponents) made a hearth in the bow of their ship, on which they rested a vessel of bronze filled with the above (ingredients) and put fire under it, and one of the crew, by means of a tube made of bronze fashioned after the manner of the toy of boys, called a squirt, played upon the enemy".

Different names are used to denote this Greek Fire. First we have "liquid fire" (pyr hygron), then "artificial fire" (pyr skeuaston), "marine fire" (pyr thalassion), "Greek fire" (pyr Romaikon), "Median fire" (pyr medikon), and finally "pyr energikon", "pyr maltakon" (soft or molten fire), "ignis volatilis" and "ignes sulfureos". It is hard to believe that these names denote the inflammable mixture made according to one recipe jealously guarded by the Byzantine state. It would seem that, as the proper chemical composition was mostly unknown to the historians and as the impure chemicals of those days made the manufacture of completely identical batches (according to one recipe) impossible, several classes of incendiary weapons were covered by the very term "Greek fire".

From the days when Vossius [2] started to investigate the early history of gunpowder, several authors have discussed the nature of Greek fire and have tried to claim that it was "solid fire" and a precursor of gunpowder [3]. Only recently a famous manufacturer of fireworks expressed this opinion as follows:

[1] Gudianus Lat., fol. 96 (Wolfenbüttel Library) (edited in Schriftenreihe der Historischen Vierteljahresschrift, Heft 1, pp. 6-7).
[2] Isaac Vossius, Variarum Observationum Liber (R. Scott, London, 1685, cap. XV).
[3] L. Lalanne, Recherches sur le feu grégeois et sur l'introduction de la poudre de canon en Europe (Paris, 1840; 2. edit. Paris, 1845).
Reinoud et Favé, Le feu grégeois et les origines de la poudre à canon (Paris, 1845).
M. Berthelot, Les compositions incendiaires dans l'Antiquité et au Moyen Age (Revue des Deux Mondes vol. 106, 1891, pp. 786-822).
C. Zenghelis, Le feu grégeois et les armes à feu des Byzantins (Byzantion, vol. VII, 1932, 1, pp. 265-286).
V. V. Arendt, Das griechische Feuer (Arch. ist. nauki i techn. vol. IX, 1936, pp. 151-204).
A. Lotz, Das Feuerwerk (Wiersemann, Leipzig, 1941).
Le col. Reyniers, Vues anciennes et nouvelles sur les origines de l'artillerie (Mémorial de l'artillerie Franç. Vol. 30, 1956, pp. 511-515).

"I was convinced of the real nature of Greek fire by a chance demonstration of something evidently very similar during trials of smoke-producing composition in 1914. The particular mixture was: saltpetre 6, sulphur 1, powdered pitch $3\frac{1}{4}$, powdered glue $\frac{1}{4}$, and plumbago $\frac{1}{4}$. This was contained in a steel mortar 3 ft. 6 in. in length and $5\frac{1}{2}$ in. internal diameter. The method of filling, which may probably have had some influence on the manner of burning, was as follows: after mixing, the mass of composition was rendered plastic by heat, taken in handfuls, moulded into balls each of which was dropped into the mortar in turn, and pressed down to fit the bore. When ignited at the muzzle, the composition burned for a time with a certain degree of violence, followed by a momentary pause. This was followed by what can be best described as a "cough", and a burning, viscous mass of partly consumed composition was blown out to a distance of upward of a hundred yards. This action was repeated with surprising regularity down the whole length of the composition".

He agrees with Hime that it was a "wet" (molten, viscous) mass of fire, made of ingredients not too difficult to obtain in Byzantium, burning with much noise and smoke and in some way connected with syphons, and he ends by saying [1]:

"Before the days of gunnery, when the longest range for a missile was that attained by an arrow from a bow or arbalest, even a hundred yards was a commanding distance from which to harass one's enemy with burning masses of adhesive fire.

With the coming of artillery and the gradual development of pyrotechny on independant lines there grew up a spirit of rivalry between the two bodies of practitioners in the use of saltpetre mixtures".

Of course we cannot deny that "gunpowder" types of mixtures, containing petroleum and saltpetre were used in Antiquity. We do not only have the testimony of the Byzantine anonym on the campaigns in the Danube provinces, but as early as the third century A.D. Julius Africanus [2] mentions a fire "which propells itself":

"The automatic fire is made in this way: grind non-burned sulphur, salt extracted from dust, thunder-stone, pyrites in equal proportions in a black mortar in the middle of the day. Add sycomore-juice and liquid Zacynthos bitumen in equal proportions until you get a flowing paste, then add a little quicklime. The mass shall be stirred with caution in the middle of the day protecting one's body, for this mixture will catch fire suddenly. Put this mixture in boxes of bronze, closed by lids and store it away from the sun's rays for their contact will set it afire".

Such explosive incendiary mixtures were often called "wild-fires" in medieval literature and they are carefully distinguished from the Greek fire type of materials squirted from flame throwers. This is clear from the lines of the epic of Richard I Coeur de Lion, written about 1300, which run:

[1] Alan St. H. Brock, A History of Fireworks (London, 1949, pp. 232-233).
[2] Julius Africanus, Kestoi VII.

"Kyng Richard, oute of hys galye
Caste wylde-fyr into the skye
And Fyr Gregeys into the see
The see brent all off fyr Gregeys".

Cheronis [1] has correctly argued that Kallinikos prepared a "liquid fire", which was ejected through tubes referred to as siphons by a "siphonator" or by a "hand-siphon" or syringe. It was also projected by means of a "strepta", an instrument which is twisted or flexible or which is discharged by turning. A roar (thunder) and smoke accompanied the ejection and it spread rapidly, burned fiercely, even on water and in some cases stank abominably.

The existence of this class of "liquid fires" has been established beyond doubt by the excellent and very complete publication of the Arabic texts on this problem and those from Crusader's documents compiled by M. Mercier [2]. We have seen that such "liquid fires" were known since 200 A.D. and there were probably several recipes, some using saltpetre, which seems to be an

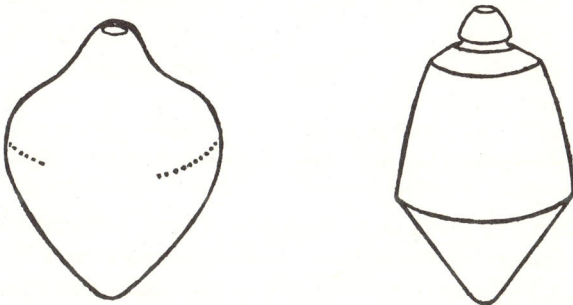

Fig. 3. The Syrian and the Egyptian (Fostat) type of incendiary bomb, the latter having a more perfect aerodynamic shape (After Mercier).

essential component of the "Greek fire" of Kallinikos. Whether this Greek fire also contained quicklime to promote self-ignition is not definitely known, but it is probable since we read in Leo's Taktika (XIX, cap. 54) that he recommends to throw onto the enemy "pots full of quicklime which when bursting spread a thick smoke which smothers and envelops the enemy in darkness". In view of the garbled versions of its composition it is impossible to establish what kind of ingredients of the calcium-phosphide type was added to some

---

[1] N. D. Cheronis, Chemical Warfare in the Middle Ages, Kallinikos' Prepared Fire (J. Chem. Education vol. XIV, 1937, pp. 360-365).

[2] M. Mercier, Le feu grégeois, les feux de guerre depuis l'Antiquité, la poudre à canon (Geuthner, Paris, 1952).
M. Mercier, Le rôle du feu grégeois dans l'Histoire coloniale (C. R. de l'Acad. des Sciences Coloniales vol. XV, 1955, pp. 185-209).

Greek fires to promote their ignition on contact with water. With Cheronis we can call Kallinikos an early pyrochemist, who produced an excellent and efficient liquid fire with a naphtha base, which also contained solid components such as saltpetre and maybe further ingredients such as quicklime. From Marcus Graecus' manuscript of the eleventh century it is clear that the Byzantines had such a "secret weapon" and that they incorporated powdered solids in their "liquid fires". They continued to use such recipes for mixtures with incendiary and explosive properties and preferred them to gunpowder in the later phases of their history. Nicetas, a witness of the fall of Byzantium reports [1]: "They threw on the houses of the unfortunates who lived near the seaboard the liquid fire which sleeps in closed pots but suddenly burst into flames and embraced the objects it reached".

This "liquid fire" was used in naval warfare and directed towards the enemy by the "siphonator" with his bronze tube, which looks like a flame-thrower in the only picture we have of such a naval battle. The word "siphon" is explained by Hesychius and by Hero of Byzantium (who wrote a Poliorketica in the tenth century) to be an engine "used to put out fires", hence a fire-engine of the type found at Bolsena and other parts of the Roman Empire and often mentioned by ancient engineers. However, the word also means "a bent tube" and "a squirt or syringe (klysteros)" or a "jet". It would seem that on board the Byzantine warships the inflammable liquid was ejected from a bronze nozzle or jet and either ignited with tinder and wicks or left to ignite itself by contact with water or maybe in some cases by the electrostatic charge incurred in ejection. The early Latin manuscript we mentioned above speaks of a "hot mixture" being ejected from a bronze "fistula" or jet and not from a pump. We have no evidence that a hand-pump of the fire engine type was carried on board the Byzantine warships. Either the siphons mentioned in the texts were hand-syringes, such as also used in land warfare, or perhaps bronze nozzles were attached to skins such as those in which the "Median oil" was traded and transported, and which could very well have served for the storage of this dangerous liquid on board the warships.

The same type of protection against such incendiaries is tried as we found in early texts. Thus a twelfth-century historian like Joannes Cinnamus tells us about the naval battles between the Byzantines and the Venetians: "The Greeks pursued them ... and tried to burn them by launching the Median fire; but the Venetians who knew their tactics and who had covered and enshrowded their vessels with linen cloth, drenched in vinegar, sailed securily".

---

[1] Nicetas Choniates, Isaac Angelos, Book I, 249.

Hand-siphons or squirts were certainly used in the army to shoot the liquid incendiary at the approaching enemy and they are more or less precursors of modern fire-arms and flame throwers.

It is not quite clear how the "strepta" worked. We have a passage by the tenth-century tactician and engineer Hero of Byzantium which runs thus:

"These stand upon the top steps and shoot fire in the face of the enemy with the hand-operated strepta, fire-arms (fire-throwing implements). And they frighten them so much, those that stand against the wall, that being unable to stand the stream of fire and the assault of battle they very quickly retire from the place. And the figures have been drawn".

From the picture given in a seventeenth-century edition of Hero [1] we get the impression that it stands closer to the modern flame thrower and fire-arm than the squirt. Feldhaus and Zenghelis have argued that they were crude hand-guns projecting a charge by an explosion, but as this seems improbable we are left with the faint indication that they were operated "by turning" to eject the incendiary ignited at the muzzle.

Then there was the pasty type of fire, the "pyr maltakon", of which Brock probably gave the correct description, with which small bronze tubes, the "cheirosiphona", were charged. These prototypes of the Roman candle were therefore fired from a prototype of the "bazooka".

There was also a class of pots, somewhat like our modern hand-grenades, which were filled with incendiary or explosive mixtures. Mercier has quite correctly pointed out that this class embraces several varieties of "pyr automaton" or "sleeping fire". Some thin-walled grenades were filled with the Greek fire mixture or with naphtha and such inflammable stuffs and flung at the enemy to be ignited by a shower of fire-arrows. Other pots with more solid walls (the Syrian and the Egyptian types which Mercier described and which differ in form only) contained inflammable mixtures lit by the soldiers who flung them at low speed at the enemy. Others may have been equipped with pre-ignited wicks drenched in sulphur and there were also hand-grenades containing the explosive "gunpowder type" of solid powder, though such pots have not been found amongst the excavated pots from Cyprus, Rhodos or Byzantium. Such pots may have contained ingredients for self-ignition, but we do not yet know sufficient about their original contents and more analyses are sorely wanted.

Such grenades were not flung by hand but propelled by the war engines in common use in Antiquity, mostly by the types not moved by the tension or torsion of sinews but by those moved by the sudden release of heavy weights,

---

[1] Veterum Mathematicum (Paris, 1693), pp. 152-153.

known to the medieval tacticians as "trebouchets". Similar engines were used by the Moslems even before the Crusades and many centuries earlier such engines were used by the Indian armies to fling stones and lead balls wrapped

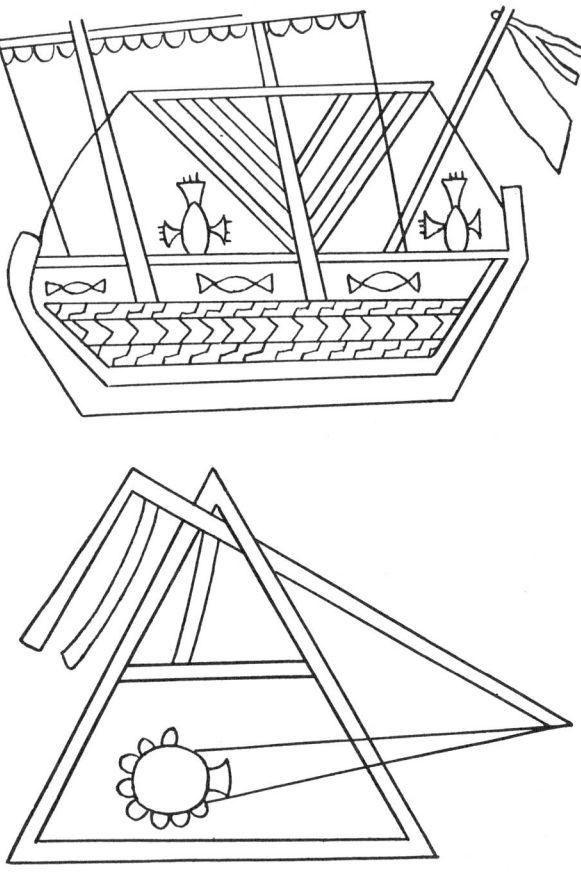

Fig. 4. Naphtha-pots on board a vessel and diagram of a machine for flinging pots with incendiary mixtures.

up in burning materials. The Mahâbhârata [1] mentions an Aśma-yantra (a stone-throwing machine) in the battle with Jarâsandha and we have further records that such engines were used in later periods to set enemy fortifications

[1] Mahâbhârata II. 42. 21.

PLATE I

Firing Greek Fire from a "strepta"
(Codex Vaticanus Graecus 1605, eleventh century).

The fleet of the Emperor Michael II (820-829) attacks the vessel of
the rebel Thomas with Greek Fire
(Mss. of Joannes Skylitzes, Bibl. Nacional, Madrid, thirteenth century).

alight [1] and that "liquid fires" containing naphtha [2] were in use in ancient India. These problems are closely related with the history of the invention and spread of gunpowder, which certainly came very early in the East, though its possible transmission (or independent invention?) to the West is yet far from clear.

Hence the "secret weapon" of the Byzantines consisted of the recipe(s) for a Greek fire of the naphtha-saltpetre type which was carefully guarded but which could not be kept secret forever. Even when Leo V the Armenian beat the Bulgarians in the battle of Mesembria (814 A.D.) he was worried by the fact that the enemy managed to get hold of 36 siphons and a considerable quantity of Greek fire. The Arabs were using "liquid fires" very early, maybe even before the battle of Cyzicus. Ibn-Djami the poet tells us that when the Caliph Haroun-al-Rashid besieged Heraclea in Cilicia (807 A.D.) "Heraclea surrendered when it saw this surprising thing, heavy engines flinging naphtha and fire. It seemed as if our fires stuck to the flanks of their citadel like stuffs dyed red hanging from the cloth-lines of the fuller". During the Crusades the Arabs handle the incendiary mixtures and the Greek fire as easily as the Byzantines though they always confess that the Byzantines had even superior methods of handling the "liquid fires", but they look down on the Crusaders who did not have such efficient weapons. The Christians from the West are constantly confounded by the ingenious ways in which the Arabs handle their "fire-arrows, pots full of sulphur, oil and all things which feed fire" as well as "liquid fires". Mercier has given us a very full picture of this chemical warfare during the Crusades, but he rightly points out, that we still do not know exactly when and where the Arabs began to use liquid fires containing naphtha, the elements of which they may have picked up in their wars against the Persians and Indians, even before they were taught their painful lesson in the Sea of Marmora.

We also know that the Arabs came to use the word "baroud" (originally saltpetre, later more generally used for gunpowder) as a synonym for "naphtha", but it still remains to be established when and where this word "baroud" came to be used in Arab texts for "liquid fire" for this will give us another pointer as to the date when salts such as saltpetre came to be incorporated in incendiary mixtures based on petroleum. We are certain that in the siege of Fostat (1168 A.D.) such mixtures containing saltpetre, sulphur and carbon were used in grenades, which recipes show that these were pasty precursors

---

[1] V. Raghavan, Yantras or Mechanical Contrivances in Ancient India (Indian Inst. of Culture, Trans. No. 10, Bangalore, 1952).
[2] P. K. Gode, The History of Fireworks in India (1400-1900) (Indian Inst. of Culture, Trans. No. 17, Bangalore, 1953).

of gunpowder. Such "gunpowder" was certainly not yet fired from a gun; in 1383 we find the word "medfaa" (cannon) for the first time. However, their explosive qualities were put to use by enclosing such "automatic fires" or "volatile fires" in the restricted space of a grenade type of vessel, which could then be cast to penetrate the enemy lines and fortifications and explode or set afire whatever it encountered. They were, for several centuries, used side by side with the true liquid fires we mentioned. The Arabs preferred these liquid fires with both incendiary and explosive properties instead of using gunpowder to propel a projectile.

The study of the story of such early liquid fires and incendiaries in the East together with that of gunpowder will reveal whether the Chinese junks, using naphtha against pirates and robbers in the Persian Gulf, learnt this secret from the Arabs or brought it there from India or China.

There is, however, no longer any doubt that the incendiary properties of petroleum and its derivatives were appreciated in Antiquity and that during Hellenism and the Roman Empire "naphtha" went to war under different guises and in varying company, culminating in the Greek fire of the Byzantines which for centuries remained a far more terrible weapon than gunpowder. Only when the explosive force of this new gunpowder was put to task to eject a projectile from a gun or cannon did the "liquid fires" withdraw to the background, but they were able to play their part again with the revival of the flame thrower in modern warfare.

CHAPTER FIVE

OIL FROM EASTERN EUROPE

When Drake drilled his first successful well at Titusville he initiated the rapid rise of a new industry based mainly on the oil of Pennsylvania during the first two decades of its career which we are discussing here. During this period new oilfields were of course discovered in the United States but between 1860 and 1880 one can call the United States a "one-crude-country", where a base material eminently suited for the needs of the young country was produced in rapidly increasing amounts. It swept away its competitors on the lamp oil market without much difficulty and the demand for other petroleum products in the United States, where industrialization had hardly begun, was much less specific than in Europe where the Industrial Revolution was now gaining momentum rapidly.

In Europe the petroleum industry had to fight its way up the ladder against fierce competition [1]. There was the cannel-coal industry of Scotland founded by James Young, the wood-tar industry, the group of manufacturers such as Hübner of Zeitz, Riebeck (Halle), Bunge and Corte (Ober-Röblingen) and Vehrigs und Söhne (Teuchern) who subjected lignite to dry distillation and who produced paraffin wax as an important byproduct, and last but not least there was the young coal-tar industry of which Julius Rütgers (Breslau) and Erkner (Berlin) were the main representatives. Against such formidable opposition the petroleum refiners of Pechelbronn, Wietze, Amiano, Galicia and Austria had to prove their merits. The petroleum industry had the benefit of the growing body of scientific knowledge accumulating in the laboratories of Europe, where the composition of crude oils was investigated and from which data conclusions could be drawn regarding the refining and testing of the various crudes and the products to be made.

In contrast to the Pennsylvanian refiner, his European colleague had to deal with a great many, widely different crudes, but this variety of base materials stimulated his ingenuity and promoted the manufacture of marketable products from all fractions of the crude and even from waste products. For even in

---

[1] H. Schwarz, Die Produkten der trockenen Destillation auf der Wiener Weltausstellung 1873 (Dinglers Polyt. J. vol. 210, 1873, pp. 205-215).

the face of this fierce competition the demand for kerosine, paraffin wax and lubricants soon became such, that local European production of crude oil proved insufficient and new sources had to be found. Though Pennsylvanian products such as Kerosine-oil of the Oleophene Co. and Paraffin from Messrs Stephenson, Philadelphia were marketed in Europe no appreciable amount of Pennsylvanian crude came to the European refineries. Some "Rangoon tar" from Burma was imported into Great Britain.

The ancient European oil-producing areas such as Wietze, Alsace and Italy could not cope with the demand. Production at Wietze declined after a short boom [1]. At Pechelbronn the extension of the galleries led to the discovery of a real oil sand in 1863 [2], the number of these "wells" increased and soon the galleries were simply used to collect the oil thus flowing from new sources. In the years 1867-1871 about 4,000 tons had still been collected yearly by the ancient method of washing the bituminous sands, this amount fell to 2,000 tons in 1872 and this production method was stopped by 1875. The oil flowing from sands into the galleries increased and it was later stimulated by the injection of water under pressure. Drilling for oil was started in 1880 and production soon reached the point of swamping the Merkwiller refinery with oil which it could not refine in such large quantities, though the volume was but a fraction of the needs of the European market.

In Italy there was more activity about 1850 and new seepages were found in the Parma-Piacenza region [3]. The Genoese Società l'Esploratrice by decree of August 22, 1866 obtained a concession with three wells at Salina and Riva in the neighbourhood of Montechino. This concession lapsed to Count Narazzani Visconti in 1878/88, then to Huber and finally in 1897 to the Società Petroli Montechino. Zipperln then started drilling with the Canadian system and obtained further concessions at Velleia (1888-1892) and at Montechiaro (1890) but serious work did not start in this region until the foundation of the A.G.I.P.

None of the old European oil fields produced sufficient oil so new supplies had to be found and were found in Galicia, Roumania and Russia. At the Vienna Exhibition of 1873 Siderow had shown Siberian crude oil, but rich resources were found nearer home in *Galicia*, which in the course of history

---

[1] R. J. Forbes, Studies in Early Petroleum History, vol. I (Leiden, 1958).

[2] Anon., Société anonyme d'exploitation minières Pechelbronn (L'Industrie du Pétrole, pp. 38-41, II. Congrès Mondial du Pétrole, Paris, 1937).

[3] E. N. Rocca, Piacenza, la città pioniera del petrolio (L'Industria Mineraria, vol. VIII, 1957, pp. 631-632).

E. N. Rocca, I pozzi petroliferi della Val Trebbia (Rivista di Diritto Minerario, 1955, 1/2, pp. 1-16).

belonged to Poland but during the period we are discussing it was partly Austrian territory.

No proper description of the Galician oilfields is given before 1721 but from Rzaczynski's book [1] it is evident that the local peasants had known several seepages for many centuries and had used the oil to grease their carts and tools and had applied it externally for several diseases. This Jesuit father inserted a special chapter on "bituminous waters" in his book which described the seepages and natural gases of the northern foothills of the Carpathian mountains. There was a cluster of seepages near Glowienka and Turszowka (in the neighbourhood of the town of Krosno) and also near Mons Admirabilis in the territory of Krakau. Timber impregnated with this oil is protected from attack by insects. The production of the seepages seems to surge and wane with the moon. In the mountains near the Russian border, near Rungury and Ropenka (Rypne) there were more seepages of the thick oil, called "ropa" locally. Many fossil shells are found in the neighbourhood. In other sites "petrolaeum" or "oleum petrae" is found, notably near Urbi Camenecensis, where it is called "lacka". A thin liquid type of petroleum is found near Drohobycz, Kopiec, Jasién and Stebnik, which is called "kipiaczka" and which is very inflammable. Rzaczynski discussed the properties of these Galician products in relation to the "pissasphaltos" of Gaebelius and Belon and the traditional bitumens described by Dioscorides.

A few further details on these seepages can be gleaned from Canon Kluk's "Encyclopaedia of the mines" (1787) and from Stanislaw Staszic's early geological descriptions of this region. After the partition of Poland the Austrian emperor Joseph II instructed the Viennese professor Balthazar Haquet to report on the mineral resources of the newly-acquired provinces, which he did after two years of exploration [2].

Production from these seepages on a larger scale started early in the nineteenth century when Joseph Hecker and Johann Mitis obtained concessions to dig shafts in the Boryslaw region and worked them from 1810-1817 [3]. Hecker managed to produce kerosine from his crude to light the nearby town of Drohobycz and petroleum lighting was also adopted by the military barracks of Sambor. Hecker had promised to deliver his lamp oil to his native town of Prague but heavy snowfalls held up the transport and forced him to pay such

---

[1] Gabriel Rzaczynski S. J., Historia Naturalis Curiosa Regni Poloniae .... (Sandomir, 1721, pp. 113-116).
[2] Balth. Hacquet's Neueste Physiologische Reisen in den Jahren 1791 und 1793.
[3] Josef Muck, Der Erdwachsbergbau (Berlin, 1903).
Sig. S. Bielski, L'Industrie de Pétrole en Pologne (Congrès Int. des Mines, de la Mét. et de la Géol., Vol. I, 1935, pp. 302-308).

heavy indemnities that he was discouraged from continuing production. In 1815 the seepages near Truskawiec, which was his most important producing shaft, gave out entirely and Hecker stopped further work in the oilfields.

Hecker's failure did not stop local production. The peasants continued collecting the thick "ropa" to be used as a cart grease and the thin "kipiaczka" which was employed to oil leather. They also used a quaint method for turning the thin crude into thicker "ropa". The crude was poured onto water in shallow holes dug in the ground and then beaten with hazel twigs to promote evaporation of the lighter fractions. Such holes were called "duczki", a name later applied to the oil seepages themselves. By 1835 there were some 20 wells sunk in the Boryslaw district alone. Concessions were given by the authorities to J. Micewski (1831) and to the Chamber of Commerce of Drohobycz (1843). Apart from the "mountain tar" many other forms of bitumens such as bituminous limestones and marls, mountain wax (ozokerite) and natural asphalt had by now been found in these regions [1].

A new start was given to the oil industry by the efforts of the Lwow pharmacist, Ignacy Lukasiewicz. In 1852 he rediscovered the proper method of distilling lamp oil from the crude and with the help of the plumber Bratkowski he constructed a good, cheap lamp for local use, which lamp was later perfected by the Viennese firm of R. Dittmar. Lukasiewicz now began to produce lamp oil systematically, first for the hospital of Lwow, where the first oil lighting was started on July 31, 1853. His production soon grew to appreciable quantities and in 1859 he was delivering 55 tons of lamp oil to the "Administration of the Northern Railways" at Vienna. Gradually he built up a complex business consisting of mines, wells, refineries and a sales company which was to flourish for several decades until his death in 1882. Robert Dom, who started producing crude oil and petroleum products about 1853, was also among the first to manufacture petroleum lubricants and naphtha from the local crude. The lubricants were sold under the trade-marks of "Mineralöl" and "Solaröl", the residue being either used as cart grease or mixed with 15-25% of sand to serve as an inferior type of paving asphalt [2].

We should, however, remember that the bulk of the crude oil was still produced from a host of small shafts sunk by peasants or private owners on their own plot. Jewish and Austrian financiers had supplied the necessary capital since 1848 and by 1853 we can speak of a real overcropping by hundreds

---

[1] H. Wachtel, Die Naphtha und deren Industrie in Ostgalizien (Dinglers Polyt. J., vol. 156, 1860, pp. 463-464).

[2] W. Jacinsky, Bergöl und Bergwachs zu Boryslaw (Oest. Z. Berg- & Hüttenw., vol. XIII, 1865, pp. 289 & 295 ff.).

Ed. Schmidt, Das Erdöl Galiziens (Wien, 1865).

L. von Neuendahl, Vorkommen des Petroleums in Galizien (Wien, 1865).

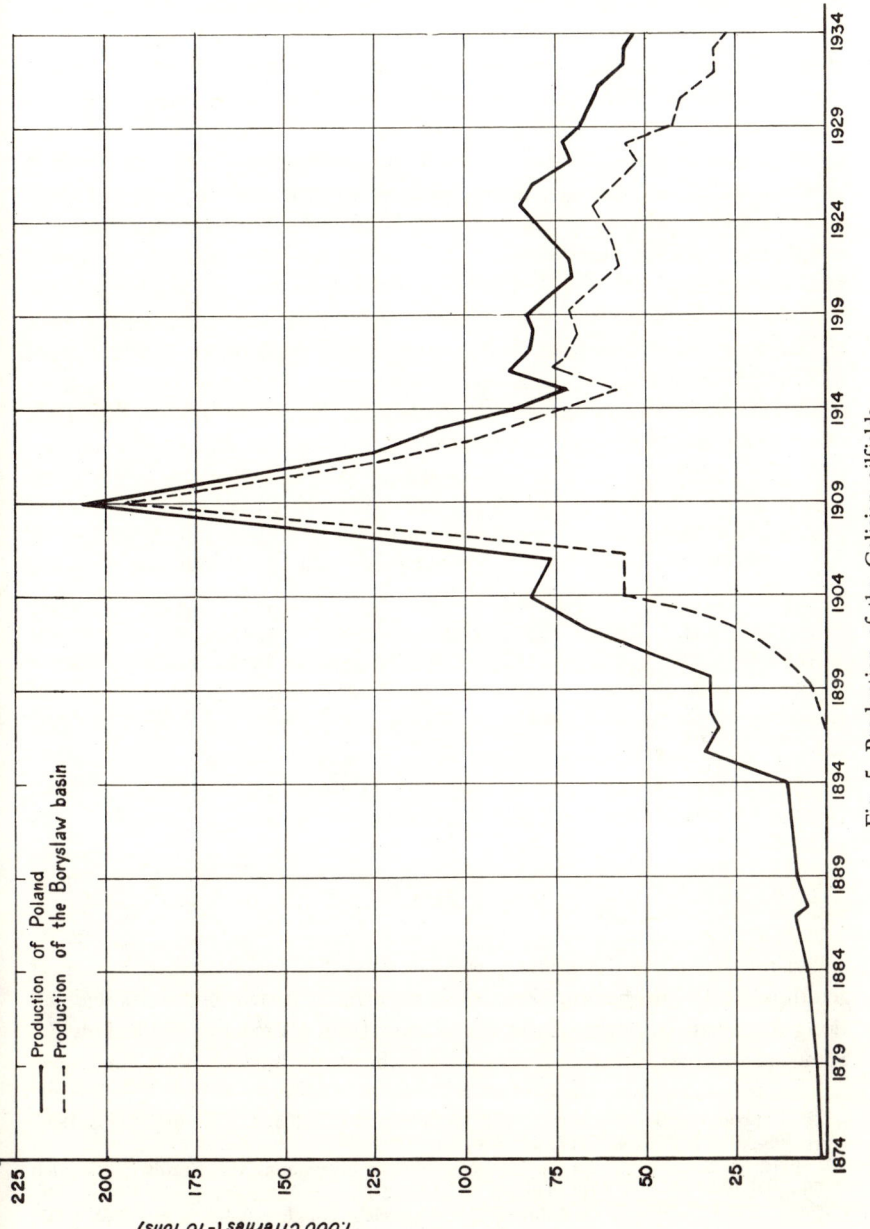

Fig. 5. Production of the Galician oilfields.

of shafts sunk in no less than 151 villages! By 1854 the demand for kerosine had grown appreciably and stimulated this local production. Two years later most wells were properly timbered and reached an average depth of 10-12 metres, producing about 500 K of crude oil per week. This handmining was very profitable for the market price was about 700 Kronen per ton and local expenses were very low. In fact the local peasants received very little for their own work and were even gradually ousted from their little holdings. Profits flowed into the pockets of the local financiers. In 1859 oil was discovered in the Wolanka district near Boryslaw and a few years later veins of "mountain wax" were being exploited, but not until 1863 were any provisions made for lighting or ventilating the shafts. These shafts were very small, the district of Boryslaw had no less than 1500 in operation by 1862, some of which were roofed-in to protect the miners against the weather and to protect the mines from theft which was rampant in these districts. There were no less than 1000 owners of these 1500 shafts and 8000 workmen were employed. The number of registered pits rose rapidly, as indicated in the table below:

Year	Shafts in Boryslaw	Wolanka	Year	Shafts in Boryslaw	Wolanka	Year	Shafts in Boryslaw	Wolanka
1865	2417	277	1878	2790	496	1890	2358	320
1866	3644	676	1879	2598	373	1891	468	304
1867	3578	818	1880	2498	334	1892	426	303
1868	3506	734	1881	2515	392	1893	403	
1869	3579	840	1882	2565	392	1894	419	
1870	3468	812	1883	2858	483	1895	449	
1871	3514	871	1884	2844	569	1896	367	
1872	3587	882	1885	2685	435	1897	404	
1873	3424	880	1886	2498	380	1898	271	
1874	3294	862	1887	2439	331	1899	194	
1875	3186	850	1888	2391	316	1900	75	
1876	2977	545	1889	2328	316			
1877	2952	527						

During the sixties the oilfields were investigated and described by expert geologists [1]. The oil-bearing formations occurred in a strip, 2-3 miles wide all along the northern slopes of the Carpathians from Moravia to the Bukowina. Some 60 different fields were known by that time but the most important

---

[1] F. Foucou, Sur le gisement de pétrole des Carpathes (C. R. Soc. Ing. Civils 1865, pp. 303-305, 317-325; 416-417; 425-426).
Bergrath v. Cotta, Über das Vorkommen und die Gewinnung des Erdöls in Galizien (Dinglers Polyt. J. vol., 181, 1866, pp. 153-154).
Anon., Über die Bergöl Gewinnung in Oesterreich (Dinglers Polyt. J., vol. 185, 1867, pp. 164-165).

region was that of Boryslaw, east of Drohobycz. In 1865 no less then 2394 shafts were in production and over 3000 abandoned altogether, producing some 9,000 Tons of oil and 4,500 Tons of ozokerite, most shafts being about 25-45 metres deep and usually deepened when production began to fall. The average "wax mine" produced 200-400 K of wax daily but some up to 3,000 K; the oil shafts produced on average 100-300 K of oil. At Boryslaw some 3,000 men were employed about this time. The industry rose rapidly and in 1874 874 firms owning 4,000 shafts employed 10,500 workmen but production was less than 7,000 Tons; by 1881 189 firms worked 1677 shafts and 1477 stood abandoned and the number of workmen had been reduced to 5063.

The production of natural wax rose quickly when in 1875 Ujhely and Pilz discovered how to manufacture ceresine from it. By 1885 no less than 1230 waggon-loads were sent to the refineries, but this local mining by smallholders declined rapidly afterwards and the production was no more than 18 waggon-loads in 1900. This decline like the temporary one in oil production was due to the exhaustion of the surface deposits, but oil production recovered when William Mac-Garvey introduced Canadian cable-drilling in the eighties and when Stanislaw Szczepanowski started to organize the industry, about the same time. Also the new mining law of 1884 promoted concentration and more efficient exploitation of the local resources. Wells were now sunk to a depth of 700 ft. and produced up to 10 Tons a day. The crude was refined by such firms as the Gallizische Aktiengesellschaft für Naphthafabrikation, Ign. Lukasiewicz (Chorkowa), Lauterbach, Goldhammer, Gartenberg und Comp. (Drohobycz), T. G. Delaval (Grybow), Dingler (Märish-Ostrau), Hichstetter and G. Wagemann [1]. Petroleum refining still remained a very profitable business, the price "loco Lemberg" fell from 320 (gold) francs to 250 only during the years 1854-1872. Many oilfields contributed to the rising production, e.g., Boryslaw, Bobrka, Palanka, Plonce, Glebokie, Wankowa, Wytrylow, Starnia, Dzwiniacz, Molodkow, Salotwina and Rybne. In 1867 there were 36 refineries treating this production, 30 of which produced lamp oil, 2 paraffin wax candles (mainly from ceresine) and 4 larger refineries which combined different aspects of the trade. The total value of the trade was said to be over 1,700,000 Austrian florins, about 1000 Tons of candles, 260 Tons of paraffin wax, 9,600 Tons of petroleum distillates (including gasoline and kerosine), 700 Tons of heavy distillates (lubricants) and 660 Tons of cart grease were marketed.

Geological investigations had revealed local outcrops of oil, even in Austria

---

[1] H. Perutz, Industrie der Mineralöle (C. Gerold, Wien, 1868, vol. I).
H. Perutz, Industrie der Mineralöle (C. Gerold, Wien, 1880, vol. II).
E. Soulié et H. Haudouin, Le pétrole (Paris, 1865).

near Gaming and Salzburg, and in Croatia (Peklenicza and Mikłoska), which oil had been analysed in a very primitive way by Winterl as far back as 1788 [1]. As drilling permitted the tapping of subterranean resources the Galician oilfields were considerably extended, the coincidence of oil seepages and salt outcrops between Wieliczka near Krakau and Wallachia near the Roumanian border had struck the geologists and refiners. Szczepanowski opened up the field of Sloboda Rungurska near the town of Kolomyja. Wolski and Odrzywolski started to drill at Bitkow and Schodnica; and the main field at Boryslaw was also discovered by drilling about 1908. Production then started to decline, notably due to the First World War, but in the early days the Galician oilfields did supply an appreciable percentage of the needs of the European oil industry and both the developments of drilling techniques and of refining methods in these regions were invaluable for the young industry.

Perutz gives us some information of the products which the Galician refiners produced, which sheds a light on the progress of proper distillation and refining techniques. In 1862 he reports that the crude oil (specific gravity 0.870) gives 6% of gasoline, 40% of slightly yellow kerosine, 30% of gasoil, 9% of lubricants and 15% of a residue containing some 3-5% of paraffin wax. The crude is transported to the refineries in wooden casks weighing from 500 to 550 K.

By 1880 the wells produce a lighter crude (specific gravity 0.835-0.847) which contains a fair amount of water. Hence this is left to settle at room temperature and the crude is then skimmed and distilled. Perutz gives the following figures for a refinery at Popieli near Boryslaw:

	Specific Gravity	*Yield in percentage*
Water	—	0.5736
Gasoline	0.650—0.750	9.3860
Lamp oil	0.750—0.850	52.4900
Paraffin oil (Wax distillate)	0.850—0.855	12.5700
Slack wax	0.860—0.885	11.4000
Heavy distillates (solid)		2.4800
Gases, coke and losses		11.1004
		100.0000

The "naphtha factory" of Przemysl obtained much better results, probably due to better apparatus:

---

[1] J. Winterl (Crell's Chem. Ann. 1788, vol. I, p. 403).

	Aver. Spec. Gravity	Yield in percentages
Water	—	2.000
Ligroin (gasoline)	—0.710	1.066
Gasoline	0.710—0.740	5.466
Kerosine (light)	0.740—0.780	9.333
Kerosine (heavy)	0.780—0.815	15.333
Solar oil	0.815—0.855	23.200
Light waxy distillate	0.855—0.880	14.141
Heavy waxy distillate	over 0.880	10.800
Residue		14.890
Gas and losses		3.771
		100.000

A certain amount of the "mountain wax" mined is of a buttery consistency, it is called "kenderbal". The light parts are separated by distillation after the wax as such has been molten and sieved to separate the earth and dirt. The hard natural wax called "wisk" is more common, it has a melting point of about 50°C. Distillation gives 30% kerosine and solar oil, 40% wax and 30% residue. This residue can be redistilled "to coke" to manufacture lubricants or it can be topped to give asphaltic bitumen of a dubious quality.

A second important source of supply was found in the oilfields of *Roumania*. The earliest texts mentioning the presence of petroleum and seepages in this region go back to the sixteenth century, that is they are a century earlier than those from Galicia. These earliest documents deal with quarrels and lawsuits between the "mochnènes", the free farmholders, and the proprietors of the oil-bearing territories, mostly members of the clergy. They discuss the licensing of the collection of oil from seepages or they are deeds of conveyance or contracts for shafts dug in the oil-bearing soil. In 1646 the monk Baudinus mentions numerous hand-dug shafts in the hills of Mosoarele, Poeni, Doftana and Pačura in the province of Bačau from which petroleum is produced.

Stoica Teodorescu in his history of the town of Câmpina mentions that "in 1697 a certain Lamba of Câmpina sold to the sons of Messire Postelnicu Vasile part of his Câmpina properties consisting of forests, waters, petroleum, etc.". When in 1716 the Moldavian prince Démetrius Cantémir published his History of Moldavia he reported that a "mineral resin" was found on the banks of the Tazlău Sărat, not far from Moinești, which was collected by the local peasantry to grease the axles of their carts. In 1729 when a piece of land is carefully delimited in an official document we read: "Having again inspected the borders towards Banești we had erected the borderstone in the waters of the Dotana and all the way to Silistea Târgului. Towards Dotana in the

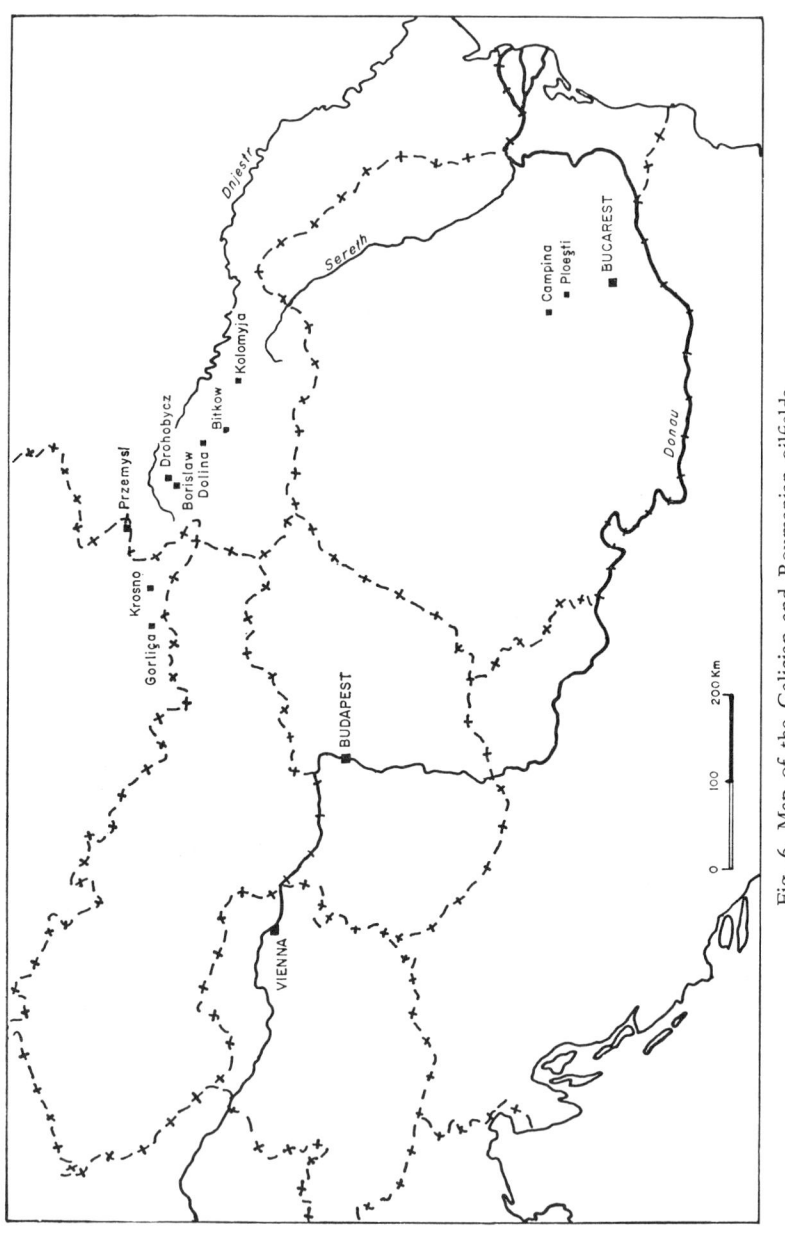

Fig. 6. Map of the Galician and Roumanian oilfields.

direction of Telega we have erected borderstone slightly above the petroleum pits". In 1745 Petcu Brasoveany lodges a complaint with the Sultan's Court at Constantinople (for Roumania was still Turkish territory) that the ladies Ilinca and Chiaina have unrightfully occupied his land (200 stânieni = 400 m. (long)) and claims his tithe of the cereals grown there and his lawful part of the 700 vedre (10,000 l.) of petroleum extracted from this territory. The Sultan's Court fixing the price of one vadra of petroleum (14 l.) at half a "thaler" fix this part at 300 vedre (4200 l.).

In the latter half of the eighteenth century Roumania was visited by several foreign scientists such as Roche, Siméon, Cara and Nagny who mention the many petroleum shafts sunk in the foothills of the mountains all over Roumania. Stephan Raicevich, a tax collector, gives us some more details on this local petroleum industry about 1780. There were two types of crude oil, a thick black oil and a reddish one. The "liquid bitumen" or "păcura" is used against cattle diseases, it lights the courts of certain castles of the bojars and it is used as a cart grease. Some wells are fairly deep notably in the region of Păcureți. Raicevich estimates the total production at 20 tons a year! His report is practically identical with that of the Count of Hauterive of 1785.

The shafts of Matița-Păcureți (Wallachia) deserve our special attention for we have evidence that the "mochnènes" of this region held special rights on the fields of Izești (Păcureți) as laid down in a document dated August 15, 1676 signed by six local inhabitants. These rights were confirmed by a charter of the prince Constantin Suțu dated July 1785 which records differences of opinion having arisen between the "mochnènes" and the monks who held the territory of Matitza. It is also clear that there were different forms of exploitation called "gropoaie", "spălătorii", "băi" and "puțuri jâmbruite", but the documents are too vague to define them properly. As an example we quote a deed of cession:

"We, the undersigned, confirm by this deed, handed to the deacon father Ioniță ot Păcureți, and of Sirghie and of Dima, that we have sold him a petroleum ditch with the workshop on the side of the hill below the well of Sârghie, full of earth which has been excavated during the last fifty years, which ditch belonged to four persons only, and we from our lawful rights give a hundred-sixty days, which is half. The agreed price is 18 Lei and the abovementioned can exploit the ditch in peace. On the sixth of November 1822.

I, Dumitru ot Ohaita   I, Ion sin Ohaita."

During the early nineteenth century many more foreign visitors from all parts of Europe visited Roumania and described the petroleum seepages. The English consul of Bucarest, W. Wilkinson, states in 1815 that the "mazout" is found near outcrops of salt, a fact that was later confirmed by the geologists and which played an important part in the exploration and production of Roumanian oil. From 1837 to 1840 the Russian scientist Demidoff, leader of a scientific mission, records the different seepages found all over the country. The wells of Păcureți, Câmpina and Baicoi were visited on several occasions by Vaillant, a teacher of the Sf. Sava college at Bucarest, who tells us that they were 20 to 30 m. deep and that they gave on the average about 100 kg. of crude oil a day. About the same time (1844) prince Bibesco who held Wallachia planned to lease the exploitation of the wells to two important Russian bankers, Trandafilof and Zeparosky but nothing came of his plans.

The modern Roumanian[1] petroleum industry was taking shape in the years 1840-1860. Very small quantities were already exported much earlier as a customs document of September 28, 1825 shows:

"Declaration number 4563. A merchant named Haim Cozac transports over the border 2500 liters of petroleum in five casks. Price Lei 2 for every decaliter = Lei 500. Taxes paid to the Customs Office of Herța, Lei 15.—".

Gradually we find small refineries buying crude oil from the miners and manufacturing lamp oil and other special petroleum products. The earliest refinery dates back to 1840. It was erected at Lucăcești (province of Bǎcau, Moldavia) by N. Choss and separated the lamp oil from the "grease", some of which was exported though most of the products were sold locally. Teodor Mehedinteanu's refinery of Râfow (near Ploesți), built in 1856, concentrated on the manufacture of a good lamp oil, then often called "lighting gas" and having obtained a concession for the supply of kerosine to the town of Bucarest he turned it into the first town completely lighted with kerosine lamps (1859). The growth of the number of refineries can be seen from the following table:

Year	Name of Refinery	Site	Capacity (tons)	Capital (Lei)
1840	N. Choss (B. Scheffler)	Lucăcești, Bacău	300	20,000
1856	Teodor Mehedințeanu	Râfov, Prahova	2,710	80,000
1857	Avram Maier (I. Grünberg)	Valea Arinilor, Bacău	873	30,000
1860	Iancu Haimsohn (Alich Leibu)	Valea Arinilor, Bacău	1,605	45,000
1860	Haim Magirescu (Hava Haimsohn)	Lucăcești, Bacău	1,323	40,000
1862	S. Nadler (Avner Grünberg)	Valea Arinilor, Bacău	2,446	50,000

[1] Mihail Pizanty, Historique de l'industrie du pétrole de Roumanie (Moniteur du Pétrole Roumain, vol. XLVIII, 1947, pp. 247-250).

Year	Name of Refinery	Site	Capacity (tons)	Capital (Lei)
1865	Avram Haimsohn (Manase Haimsohn)	Valea Arinilor, Bacău	1,468	50,000
1865	Matache Nicolau Dr. Aisinman "Standard"	Ploești, Prahova	5,270	300,000
1868	Nathan Epstein	Valea Arinilor, Bacău	20,000	300,000
1870	Costache Manole	Ploești, Prahova	19,608	200,000
1872	H. și S. Bernstein	Lucăcești, Bacău	700	15,000
1872	Matache Gogulescu	Ploești, Prahova	5,488	500,000
1878	Gr. N. Grigorescu	Ulmi, Dâmbovița	4,057	80,000
1879	David A. Grünberg	Valea Arinilor, Bacău	1,525	50,000
1880	M. I. Schapira ("Astra" C. M. Pleyte Mz.)	Ploești, Prahova	44,890	1,043,000
1880	N. Zilberman (Alter Schwartz)	Onești, Bacău	700	15,000
1882	M. Anastasiu (H. Goldenstein)	Simileasca, Buzău	3,154	17,000
1882	"Petrolul Buzău"	Buzău	1,000	25,000
1883	N. Constantinescu	Podenii-Vechi, Prahova	273	12,000
1883	'Uranus"	Ploești, Prahova	1,400	30,000
1884	Sc. arlat Parskeva și Dobrescu	Ploești, Prahova	15,125	450,000
1884	I. Dumitrescu (N. V. Ciocărdel)	Păcureți Prahova	625	10,000
1885	G. Constantinescu	Târgoviște, Dâmbovița	1,939	50,000
1885	Poenaru & Co. ("Pallas")	Galați, Covurlui	800	20,000
1886	N. Bălășescu (Zaharia Panțu)	Brănești, Dâmbovița	659	30,000
1887	Blum & Fainaru	Mărgineni, Bacău	1,859	20,000
1888	Isac Haiman "Victoria" (Petrocarbon)	Colanul, Dâmbovița	3,396	100,000
1888	Al. M. Ghiuță	Măgurele, Prahova	307	10,000
1888	A. Birnberg ("Progresul")	Tătărani, Prahova	1,402	31,000
1889	Ion Grigorescu	Colanu, Dâmbovița	16,608	582,000
1889	Soc. rom. pentru ind. petrolului	Militari (București) Ilfov	30,000	100,000
1890	Soc. Olandeză (Petrol-Latina)	Plopeni, Prahova	143,720	1,600,000
1891	Al. C. Gheorghiu (Coroana)	Mizil, Buzău	521	40,000
1894	Câmpeanu et Co.	Colanul, Dâmbovița	38,400	1,000,000
1894	A. Haimsohn	Costișa, Neamț	1,348	30,000
1894	Moses Frischhoff (Danube Oil)	Valea Arinilor, Bacău	3,875	7,000

Year	Name of Refinery	Site	Capacity (tons)	Capital (Lei)
1894	Socolescu și Th. Rucăreanu (Lumina)	Ploești, Prahova	2,228	50,000
1895	Leon A. Leibu	Adjud, Putna	1,000	20,000
1897	Steaua Română	Câmpina, Prahova	516,890	8,500,000
1898	"Noris"	Ploești, Prahova	2,000	50,000
1898	Soc. Română și Soc. Olandeza (Aurora)	Băicoi, Prahova	275,520	7,680,000
1898	Niculescu-Ciufu	Ulmi-Târgoviște, Dâmbovița	789	3,370,000
1898	V. H. Stinghie	Mărăcineni, Buzău	1,528	25,000
1899	Ionescu et Lecordier (Baneasa)	Băneasa, Ilfov	26,864	500,000
1899	N. Sandu et M. Frischhoff (M. Munteanu)	Onești, Bacău	5,822	90,000
1900	"Trajan" (Colombia)	Cernavodă, Constanța	146,080	2,550,000
1900	"Helios" (Apollon Petrol)	Târgoviște	1,800	30,000
1901	Melic Șetian	Băicoi, Prahova	1,708	50,000
1902	A. Niculescu	Matița, Prahova	215	40,000
1904	Româno Americană	Teleajen, Prahova	154,680	5,000,000
1904	"Venus" (Astramina)	R Sărat	2,000	30,000
1904	Grünberg și Veinstein	Luncani Bacău	400	12,000
1905	C. Haimsohn	Buhuși, Neamț	250	8,000
1905	Vega (Concordia)	Ploești, Prahova	283,260	5,000,000
1906	Aquila Franco-Română (Colombia)	Ploești, Prahova	141,090	2,100,000
1907	Dr P. Manea	Luncani, Bacău	575	10,000
1907	Gardner (Dr. Smarack)	Roznov, Neamț	500	10,000

The small local refineries gradually made way for larger firms and the concentration of the oil business led to the foundation of the following important petroleum firms during the early period (1860-1880) which we are discussing:

Firm	Founded	Capital (Lei)	Scene of Activity
Th. Mehedințeanu	1857	2,000,000	Matița, Apostolache, Ochișori
Valachian Co. Ltd. (Jackson Braun et Co.)	1868	7,000,000	Țintea
Grigore Monteoru	1865	2,000,000	Sărata-Monteoru, S"rani Ganea
Joseph Therler	1880	1,000,000	Tazlăul Sărat

Firm	Founded	Capital (Lei)	Scene of Activity
Suchard & Co.	1880	2,000,000	Colibași
Hernia Kornhauser et Co.	1880	1,000,000	Câmpina
Thoitz et Co.	1880	1,500,000	Doftănet
Principele Știrbey	1883	500,000	Voila
Ch. Stroe	1884	300,000	Buștenari, Tega
N. Iacob	1885	200,000	Coculești-Măguricea
G. Gr. Cantacuzino	1885	800,000	Drăgăneasa
Hildebrand & Co.	1886	1,250,000	Popești, Matița
I. Grigorescu	1886	1,000,000	Glodeni, Colibași, Reșca
Ofenheim, Singer et Co.	1886	1,000,000	Câmpina
"Soc. Rom. p. Industria petrolului"	1889	4,000,000	Solonti, Moinești

The crude oil was still obtained from hand-dug shafts for modern petroleum drilling methods did not reach Roumania before 1882 from Galicia, when the Canadian cable-drilling system was introduced in this country. Still thousands of shafts were exploited over an area of 230,000 hectares, most of them being about 50 to 70 m. deep but those of Moinești reached a depth of 120 metres. The landlords exacted a rent of one third of the oil obtained. The cost of production was very low and amounted to about two gold francs per 100 k., but the bottleneck was the primitive and expensive transport to the refineries. The railway system was still in an early stage of development and the casks of oil often had to travel some 20 miles to reach the nearest station. Even the most important refineries concentrated around the town of Ploești had to get their oil in carts from a distance of 10 to 15 miles inland. Contacts by rail with Bucarest were excellent but the connections by rail to Kronstadt (and hence to Hungary and Austria) and to the sea were still under construction and held up the development of the young industry. Still exports (often by water) from Roumania went to Budapest, Vienna and Odessa. The heavy crude was split by distillation into lamp oil and residue, the latter being sold locally as a cheap fuel. The lighter crude was distilled into:

Prime kerosine	40%
Kerosine, second quality	20%
Waxy distillate (for paraffin wax)	22.5%
Fuel and Losses	17.5%
	100.0%

The improved railway system proved a boon for Roumania for its petroleum products could be exported tax-free to the Austrian monarchy whereas American petroleum products were taxed at $3\frac{1}{2}$ gold francs per 100 kg. In 1876 the total production amounted to about 7000 barrels, 2350 of which were exported to Central Europe.

The story of the Caucasian oilfields goes back many more centuries [1] Mirzoeff had built the first well-equipped refinery at Sourachany in 1863 and ten years later there were 23 refineries at Baku. Large amounts of oil were obtained on the Apscheron peninsula, even the gases being sufficient to provide enough fuel for the refineries and factories. Other fields were equally productive, the small field of Anapa in the Kuban area producing 10,000 to 12,000 buckets of oil daily in 1866. Ragosine had started making lubricants from mazout and this liquid fuel was growing more and more important in Russian shipping and land transport. The growth of the Baku petroleum industry was closely bound up with better communications with the Black Sea region and the building of the Transcaucasian railway, pipelines and tankers. These developments, which have been ably described elsewhere [2], enabled the industry to refine 200,000 tons of crude in 1872 and this amount increased rapidly. The specific gravity of the crude oils ranged from 0.860 to 0.875, the yield in kerosine being 30-35% (spec. grav. 0.820). Lissenko pointed out [3] that Russian kerosine, with the same boiling range as American kerosine, had a higher specific gravity. He also was able to transform the vapours of Baku oil, led through a tube filled with glowing charcoal, into aromatics such as benzole, naphthaline and anthracene, which he could identify, though he did not succeed in preparing a single individual hydrocarbon from the Baku crude itself. The main reason why Caucasian petroleum products were not yet important on the Central European market was the high cost of transport which amounted to over one ruble for every Pud (= 16.3 kg.) of oil shipped from Baku to Vienna, prohibitive costs if compared with the few kopeks for which this amount of crude was delivered to the refineries of Baku. Hence, however important Russian oil production was to be for Europe in the years to come, it played only a minor part in 1880.

In this period the crude oil production of Europe rose mainly through the contributions of Galicia and Roumania. It amounted to:

---

[1] R. J. Forbes, Studies in Early Petroleum History I (Leiden, 1958, pp. 154-162).
[2] F. C. Gerretson, History of the Royal Dutch (Leiden, 1957, vols. I & II).
[3] R. Lissenko, Über russisches Kerosin (Ber. Dtsche Chem. Ges. 1878, pp. 341 ff.).

Crude oil production of Europe (1860-1880)

Year	Barrels	Year	Barrels
1860	36,240	1871	278,230
1861	44,930	1872	291,370
1862	50,630	1873	610.570
1863	70,260	1874	883,190
1864	97,780	1875	1,238,560
1865	108,300	1876	1,671,270
1866	138,260	1877	2,131,500
1867	176,150	1878	2,728,750
1868	266,490	1879	3,271,870
1869	364,450	1880	2,919,580
1870	324,690		

The bulk of this production no doubt came from Galicia, the other fields in Wietze, Pechelbronn, Amiano and Roumania contributing only relatively small quantities with Russia only looming up as a possible contributor to the European market. Still the great variety of crudes stimulated the European refiners to find markets for a number of petroleum products which they manufactured by newly discovered methods.

CHAPTER SIX

OIL FOR MILLIONS OF LAMPS

The nineteenth century produced the two main factors in the rise of the young petroleum industry, new lamps and a new type of lamp oil.

The new lamps.

Up to the days of the French Revolution candles or oil lamps were the only means of lighting the home The oil lamp was nothing but a bowl with a spout in which a wick barely sucked in sufficient amounts of olive oil (or other vegetable oils) to feed a weak, unstable and smoky flame. Some of these lamps had a central wick supported by some kind of wick-holder. In the middle of the eighteenth century the crusie became very popular, an oil lamp, which was placed in a second bowl to collect the drippings of the wick.

The new oil lamp was a creation of Argand, whose career is typical of the inventor of the early Industrial Revolution, Francois Pierre Ami Argand was born at Geneva on July 5th, 1750 as the son of a famous watchmaker. His father wanted him to take orders but the lasting impression, which the lectures of that famous Geneva physicist H. B. de Saussure made on him, changed his fate and provided with a letter by de Saussure addressed to the well-known French chemists Lavoisier and Fourcroy he went to Paris in 1776.

In Paris he soon attracted attention by the important papers which he sent in to the Académie Royale des Sciences. He described projects to improve the stills used to manufacture alcohol from wine and this drew the attention of the vintners and alcohol manufactures of Southern France. A certain Monsieur de Joubert asked him to move to Montpellier, which ancient city had a very active university, which specialized in what we now call chemical engineering problems, one of which was the economy of alcohol production from wine. At Lyons, travelling down to Montpellier, Argand met Joseph Montgolfier, son of a paper manufacturer, who with his brother Etienne was planning to realize travel by air with the help of balloons.

The casual meeting with the Montgolfiers developed into a life-long friendship. At Lyons the young men enthusiastically discussed their various plans. One of these was of prime importance, for the lift of the balloons was to be provided by hot air and how could they produce this better than with the burner which Argand had invented. The years between 1770 and 1785 were of prime importance for the development of modern chemistry. In 1775 Priestley had isolated and described oxygen but only five years later did Lavoisier recognize its part in all combustion reactions. Argand shared the interest which the scientists of those days took in all combustion reactions. He rightly considered the smoke emitted by the primitive oil lamps of the eighteenth century as the effect of insufficient oxygen supply to the flame. He held that the light was produced by oxidation of the oil in the wick with the help of the surrounding air. In this process glowing particles of carbon were formed which burned up in the outer mantle of the flame. Hence an air supply which was both sufficient and adjustable was essential. Argand believed that he could realize this by using a cylindrical wick with a central air supply, which wick was to be adjustable by a mechanism to turn up the wick. By placing a short metal cylinder on a perforated disc attached to the wick-holder an air supply to the outer mantle of the flame would be achieved, which would lead to a stable combustion of the carbon particles in the luminous flame and eliminate smoke.

These correct principles seem to have dawned on Argand in 1780, he discussed them with the Montgolfiers and demonstrated some of his lamps at a meeting of the States of Languedoc at Montpellier in 1782. In the same year he travelled back to Paris with the Montgolfiers, where he helped them during the demonstrations of their balloons to the Court at Versailles, together with several other scientists. Here he came to meet Antoine Arnold Quinquet, an apothecary of the Rue du Marché aux Poirées, and Amboise Bonaventura Lange, a grocer and a clever businessman; both men were to play a fateful part in Argand's life.

There is no doubt that Quinquet and Lange often heard Argand discuss his new oil burner, which was to turn the oil lamp into a real source of light, and they must have seen many of his demonstration models. Argand never made a secret of his plans and inventions and he spoke of them frequently for he was trying to interest many authorities and bankers in this new invention. This was the reason why he crossed the Channel in 1783 to visit England, the leading technical nation of those days. De Luc had managed to introduce him at the court at Windsor where he demonstrated his lamps to King George III. From his letter to Montgolfier of November 29, 1783, the king had advised him to take out a patent for his oil lamp. They had interested him much more than the balloons filled with hydrogen. Indeed Argand soon obtained a British patent (No. 1425 of July 3, 1784) and he succeeded in interesting the firm of Boulton and Watt, the leading engineers and machine-builders of Great

Britain, in his lamp. Their factory at Soho (Birmingham) started producing them in 1785.

When Argand returned to Paris that year he had the shock of his life. The beautiful new lamps of Quinquet and Lange were the talk of the town, their success being due to the demonstration in the Comédie Française during a performance of "The Marriage of Figaro" by Beaumarchais on April 27, 1784! Quinquet, a pupil of the Abbé Baumé, and Lange had, according to their propaganda, "improved the lamp of a stranger by adding a small cylinder of crystal glass enclosing the flame, which enhanced the effect of the cylindrical wick and the central air supply to the wick". This certainly added to a proper draught of air along the outer mantle of the flame and improved its adjustability. Argand's British patent does not mention this long, transparent lamp-glass anywhere. With Meursier and others he simply added a short metal cylinder which covered only part of the outer mantle of the flame. Quinquet was indeed the inventor of the lamp-glass which, combined with Argand's burner, yields a satisfactory lamp.

Recovered from this shock Argand began a strenuous paper war with Quinquet and Lange, accusing them, and partly he was right, of having stolen his inventions. He tried to win over several authorities to his side to help him in establishing his rights. Unfortunately Argand had never completed his lamps down to the smallest details having primarily been interested in this lamp as a source of hot air for the Montgolfier balloon rather than as a source of light. In any case Lange without mentioning Argand's name had presented the Argand lamp with a lamp-glass to the Académie on February 21st. Argand tried to explain that he had gone to Great Britain, because nobody in France could make lamp-glasses from the newly-discovered flint glass, but this did not help. Lange, the best businessman of these three had worked hard to make the Argand burner and lamp-glass a success and he had managed to get the manufacture of his "angel's lamp" (lampe angélique) on its feet. Hence, little choice was left to Argand but to come to an understanding with Lange and together they managed to obtain a privilege for the manufacture of their lamps from the Minister de Calonne on January 5, 1787. Quinquet, however was rightly wary of Lange's business methods and he had started making the lamps himself, selling them in his pharmacy. He continued the paper war to defend his invention of the lamp-glass.

For a short while fortune seemed to smile on Argand who, with the help of his influential friends, managed to found a joint-stock company for the manufacture and sale of his lamps. A hundred shares of 10,000 livres each were to be issued, but the French Revolution intervened. The only person who derived some profit from the new lamps was Lange who, we must admit, improved it on several points. Thus he introduced the constriction in the lamp-glass, which improved its action as a chimney.

Argand, a child in money affairs, was driven to despair by Lange's manipulations. He was also threatened by various manufacturers, who owing to the very primitive patent laws of those days, went on manufacturing his lamps in Great Britain and France disregarding his patents. His correspondence is filled with complaints against "the scoundrel Ramsden", Lange and others-but he was no match for them. The processes he started against these competitors ruined him financially. He still worked at many new inventions such as a hydraulic ram, which he was to sell together with Montgolfier, but all his attempts failed. Financial ruin seemed certain and in vain he tried his luck at the occult sciences and alchemist recipes for the making of gold. On October 14, 1803 he died ruined and disillusioned, in the same year as his former friend and collaborator Quinquet who, less ambitious, managed to lead a safe existence in a well-run pharmacy selling his lamps as a side-line. The only one of the three pioneers of the modern oil lamp who managed to assemble a reasonable fortune was Lange, the business man of this strange trio.

However, the new lamp was still far from perfect. The common fuels of those days were colza oil and patent oil (rapeseed oil), their lack of fluidity limited the amount of oil which the wick could suck up by capillarity and the result was a fairly small flame and little light. True, the eighteenth century had tried to cope with this evil by introducing the bird-fountain type of feed and placed the reservoir above the wick, thus avoiding the constant filling of the lamp in order to have sufficient luminosity. The early Argand lamps too had a high annular reservoir which was heated by the combustion gases and thus made the oil flow more freely to the wick. However, such constructions were dangerous as they affected the stability of the lamp and limited the light to a cone below the reservoir.

Many attempts were made to increase the supply of oil to the wick [1]. Meister Grosse of Meissen (Saxony) had applied pressure to the surface of the liquid by means of s small hand-pump in 1765. The Sinumbra ("shadowless") lamp of the Parisian watchmaker Carcel was a better solution. He applied pressure to the surface of the liquid by means of a spring mechanism, but this was an expensive solution and such lamps could only be bought by the very rich. In 1836 Franchot invented the "moderateur" or "regulator" lamp, which was

[1] F. W. Robins, The story of the lamp and the candle (Oxford University Press, London, 1939).

Werner Bloch, Vom Kienspan bis zum künstlichen Tageslichte (Dieck & Co. Stuttgart, 1925).

W. T. O'Dea, The Social History of Lighting (Routledge and Kegan Paul, London, 1958).

further improved by the German Neuburger in 1854. The reservoir of this type of lamp contained a piston which was forced down by a spring and thus exercised pressure on the surface of the liquid, increasing the supply of oil to the wick. The oil was forced to the wick through a small tube containing a valve which allowed adjustment of the oil supply. Such lamps burned some eight or ten hours before they had to be refilled.

Then there was the common Veritas lamp, a direct descendant of the Argand lamp and in 1860 the Duplex lamp was introduced, which did not contain a round burner but two parallel holders for flat wicks. Hence we have three types of lamps, those with solid wicks; with flat wicks and those with a circular wick with central air supply, the latter still giving the largest amount of light for a given oil consumption.

The new fuels

The new lamps however, could never supplant the candles unless a better fuel was found. This fuel still consisted of fatty oils. In southern countries olive oil or sesame oil were burnend, in northern countries colza oil or patent oil (rapeseed oil) but also whale oil and seal oil or fish oil. What strikes us in this series of lamp oils is that they were all fit (or could be made suitable) for human consumption and that therefore the lamps of the early nineteenth century were burning food which is now turned to better use since the advent of coal oil and kerosine.

The new coal oil was the final result of scientific work starting with the gasification and dry distillation of various materials by Lebon. Reichenbach, Selligue and many others submitted bituminous substances to dry distillation [1]. Reece worked on the gasification of peat in the laboratory of Pelouze at Paris in 1849. However, the man who turned these experiments into a thriving business was James Young whose patent dates back to 1850 [2]. He had started preparing a lamp oil from the crude of a seepage in a coal mine at Riddings (Yorkshire) in December 1847. Later he turned his attention to the "cannelcoal of Boghead near Bathgate" from which he produced his "paraffine oil" (lamp oil) and "paraffine" (wax).

We have an accurate description of the process used by Young, written by C. Greville Williams for Ure's Dictionary: "NAPHTHA, Boghead or Bathgate. Syn. Photogen, Paraffine Oil. For several years a naphtha has existed in commerce under the above name. It is now prepared on an immense scale

[1] N. B. Delahaye, L'histoire des schistes bitumineux (Revue Sci. et Industr., vol. XXXVIII, 1850, page 49).
R. J. Forbes, Studies in Early Petroleum History I (Leiden, 1958, ch. XIII).

[2] James Young, Brit. Patent of October 17, 1850 "For improvements in the treatment of certain bituminous substances mineral, and in obtaining products therefrom (parafine oil and parafine) (see Mechanics' Magazine vol. LIV, 1851, pags. 334-336).

PLATE II

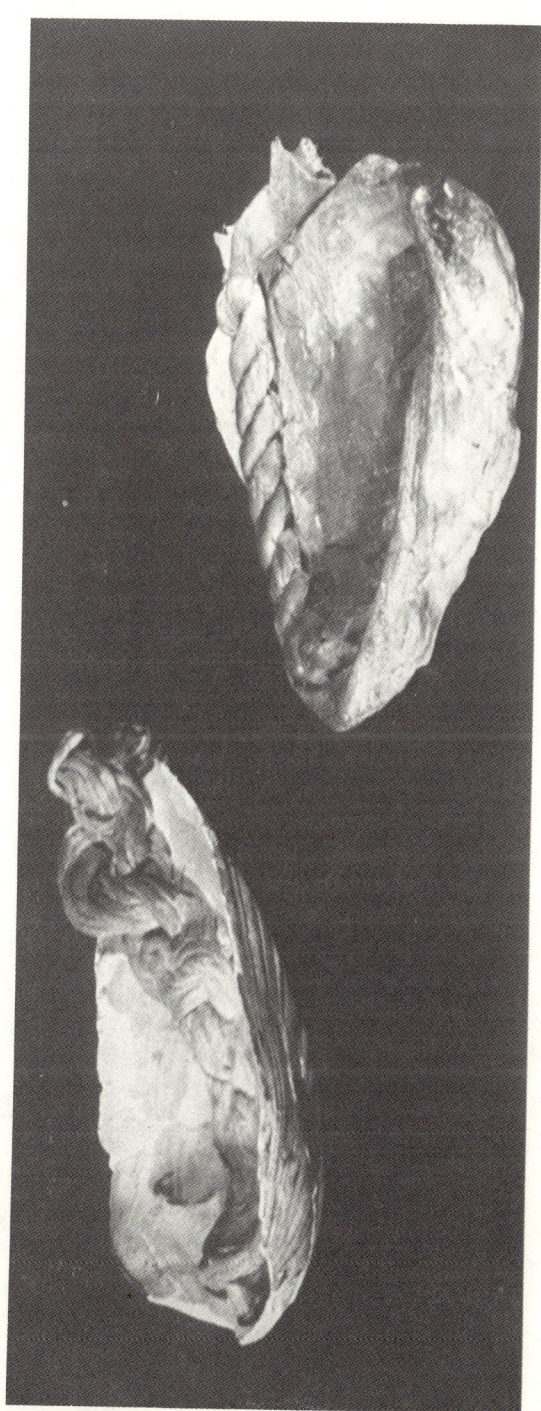

Primitive "shell" lamps still in use in Cornwall in the nineteenth century (Photo Science Museum, London).

PLATE III

Moderateur lamp of the early ninieteenth century.

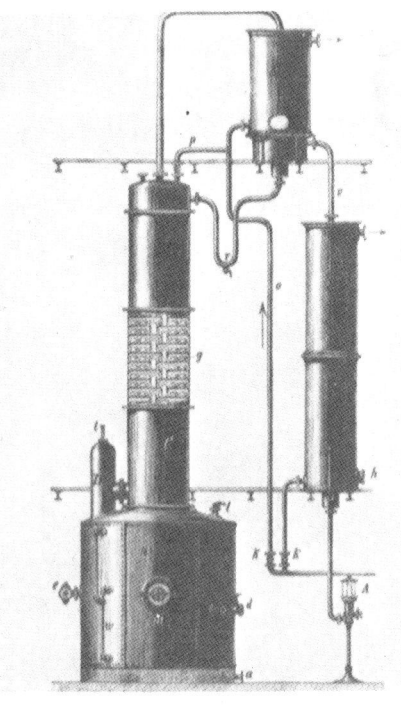

Heckmann vertical still with fractionating column used for the redistillation of light petroleum products around 1880.

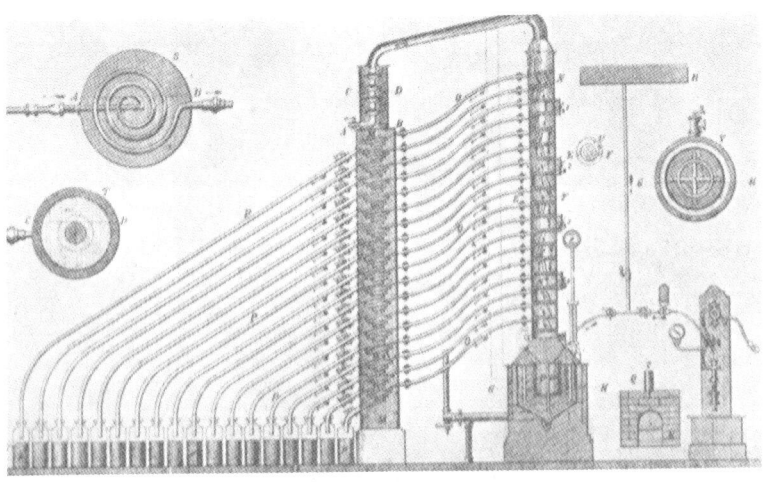

The Coffey still.

in various parts of the Old and New World. It was, we believe, at first procured solely by the distillation, at as low a temperature as possible, of the Torbanehill mineral or Boghead coal, but now it has been ascertained that any cannel coal, or even bituminous shale, if subjected to the same treatment, will yield identical products.

Photogen may be recognised at once by its low specific gravity, the ordinary kinds (boiling between 290° and 480°) having a density of about 0.750; whereas coal naphtha cannot be brought by any number of rectifications below 0.850.

The less volatile portions of the first runnings of photogen, contain a considerable quantity of paraffine, so much so indeed, that the oil is extensively used, under the name of paraffine oil, for lubricating machinery. A mixture of the more and less volatile portions is employed for burning.

Preparation of crude paraffine oil.

The following is an outline of the process employed by Mr. James Young. The best coals for the purpose are parrot, cannel, and gas coals, and especially the Boghead coal. It is well known that the latter yields a very large quantity of ash or earthy residuum, when burned in an open fire or distilled: this does not, however, interfere in the least with its value as a source of photogen. It is convenient, previous to placing the coals in the still, to break them into fragments of the size of hen's eggs, this operation enabling the heat to penetrate more readily throughout the mass. The apparatus for distillation merely consists of an ordinary gas retort, from the upper side of which a conduction pipe passes to the condensing arrangement. The latter must be moderately capacious, and not kept cooler than 55° F. The reason of this is, that if too small or too cool, the paraffine is liable to accumulate and choke up the exit pipe. When the retort has been closed in the ordinary manner, it is to be heated to a low red, but not higher, until no more volatile products distil over. If the heat rises above the temperature indicated, a considerable loss is incurred, owing to formation of too large a quantity of olefiant and other gases. The retort must be allowed to cool down considerably before the insertion of a fresh charge, otherwise much is lost before the joints are made tight.

Mr Young states that instead of driving over the whole of the fluid by distillation in the manner described, a portion may be conveyed at once from the still by having an opening in its lower part communicating with a pipe passing to some convenient recipient. By this arrangement, the products from the coal are removed from the still the moment they have assumed the liquid form. It is preferable, however, in almost all cases, to distil the hydrocarbons over in the manner first mentioned.

The product of the operation conducted as above is crude paraffine oil. It will sometimes begin to deposit paraffine when the temperature has only fallen to 40°. During distillation a certain quantity of gas is necessarily produced, but it is essential to economical working that the amount should be as small as possible. To effect this, care must be taken not only to use as low a temperature as is consistent with the distillation of the oil, but also to apply the heat gradually and steadily.

Purification of the crude paraffine oil for lubricating purposes

The oil is run into a tank and heated by a steam pipe to about 150° F. This causes the water and mechanically suspended impurities to separate. The fluid should be permitted to repose for about twelve hours before being run off. The impurities and water (owing to their being specifically heavier than the paraffine oil) remain at the bottom of the settling tank.

The crude oil, after separation of the mechanically suspended impurities, is then to be distilled in an iron still attached to a condenser, kept at a temperature of 55°, with the precautions to prevent choking up which were previously described. The distillation is conducted by the naked fire, until no more can be driven over. The dry coke-like mass which remains in the still is to be removed before making a fresh distillation.

To each 100 gallons of this distillate, 10 gallons of commercial oil of vitriol are to be added, and the mixture is to be well mixed for about one hour. The apparatus best adapted for this admixture is described in the article *Naptha (Coal)*. After the thorough incorporation of the oil and acid, the whole is to be allowed to rest for about 12 hours, to enable the acid "sludge" to sink to the bottom of the vessel. The fluid is then to be run off into another vessel (preferably of iron); and to each 100 gallons, 4 gallons of caustic soda, of the specific gravity 1.300, is to be added. The soda and oil are then to be well incorporated by agitation for an hour, so as to thoroughly neutralise any acid which has not settled out, and also to remove certain impurities which are capable of combining with it.

The oil so purified, is a mixture of various fluid hydrocarbons, to be presently described, holding in solution a considerable quantity of paraffine. The more volatile hydrocarbons may be removed by the following process:

The purified paraffine oil is to be placed in an iron still, connected with a condensing arrangement. The still is then to have run into it a quantity of water, about equal to half the bulk of the oil, and this distillation is to be continued for 12 hours. It is obvious that a great portion of the water would distil over, if not replaced during the progress of the distillation. It is preferable to perform the distillation by means of direct steam. A volatile clear fluid will distil over with the water. The naphtha so procured is lighter than water, and soon separates from it. It contains little or no paraffine. The oil remaining in the still is, of course, richer in paraffine by the amount of naphtha removed, and the separation of the solid hydrocarbon is facilitated greatly by the process. The naphtha which distils over with the water in the above process, is the fluid, the chemical nature of which is fully described in this article. A very volatile spirit may be extracted from it, by rectifying it in the apparatus recommended for benzole.

The further purification of the paraffine oil is managed as follows: —After separation from the water it is run off into a leaden vessel, and 2 gallons of sulphuric acid added for each 100 gallons of oil. The mixture is to be well incorporated for 6 or 8 hours, after which it is allowed to remain quiet for 24 hours, in order that the acid and any combined impurities may settle to the bottom of the tank. The oil is then to be carefully run off into another tank, and to each 100 gallons 28 lbs. of chalk ground with water to a thin paste are to be added. The whole is to be mixed together until every trace of sulphurous acid is removed, and is then kept at about 100° for a week, to permit impurities to settle. The oil thus prepared is fit for lubricating purposes, either per se or mixed with an animal or vegetable oil."

It was Young's American competitor Abraham Gesner (1797-1864) who coined the word "kerosene" which stuck not only to the lamp oils prepared from bituminous coals and the like but also to the lamp oil distilled from crude oil.[1] Kerosine rapidly supplanted the older fatty oils and even coal oil in the lamp oil market. When Prof. Benjamin Silliman had analysed the samples which Elvereth and Bissell had sent him he had immediately found that lamp oil prepared from crude petroleum could give more light-units per unit of weight than any of the fatty oils. This result of his photometric investigations of 1855 struck the layman even without the help of scientific instruments by simply comparing the results in his lamp at home.

The task of the early refiners was limited to the production of as large a proportion of the crude oil as could be sold as lamp oil. Evidently the heavier fractions had to be eliminated and the very lightest could not be used without grave danger to the consumer. Distillation and fractionation therefore claimed their attention. Oppler[2] in his handbook describes the earliest phase of European petroleum refining. He points out that distillation had been pro-

[1] K. F. Beaton, Dr. Gesner's Kerosene (The Business History Review, March 1955, pages 28-63).
[2] Theodor Oppler, Handbuch der Fabrikation mineralischer Oele (J. Springer, Berlin, 1862).
F. Rossmässler, Paraffin- und Solaröl Fabrikation (Wagners Berichte 1862, page 685; Ill. Gewerbe Ztg. vol. II, 1860, page 88).

moted by the research work of Selligue, Mansfield (1847), Young, Wagemann, Vohl, Brooman and Barry, most of whom worked with oils obtained by dry distillation of various base-materials. Since then Bancroft and Warren de la Rue had distilled Burmese crude oil, which they often called "Rangoon tar", using superheated steam for the heavy fractions. Oppler realizes that different refining processes should be used for products of dry distillation and those obtained from crudes. He points out that over one hundred patents have been taken out for the manufacture and refining of lamp oils since Dundonald (1781), but that most of them now deal with kerosine.

Generally speaking Oppler advises refining the products of dry distillation with a strong sulphuric acid, followed by an alkali wash, but in most cases a final redistillation is necessary. Refining kerosines from petroleum is much simpler. Steam of different temperatures can be used to separate the different fractions of the crude. In general kerosines with a specific gravity below 0.828 need not be refined beyond the distilling stage and can therefore be sold as "straight-run" kerosine. Heavier oil fractions can be redistilled into a light lamp-oil fraction (which needs but a simple wash with a caustic soda solution of 1.400 specific gravity) and the heavier parts, which after a proper acid wash: alkali treatment and redistillation, are excellent lubricants. He prefers kerosines to the lamp oils prepared from fatty oils, as they never become rancid, improve with age, though they never lose their smell which is due to small amounts of sulphur compounds.

Collecting two bulk-distillates from the crude oil was general practice in Europe during the sixties [1], the lightest parts containing all distillates with a specific gravity up to 0.828-0.830 and the other fraction embracing the heavier oils. Some refiners sold the light distillate as such but most redistilled it, collecting all fractions up to 0.735 in the "naphtha" tanks, those from 0.735 to 0.820 as kerosine and added the residue to the lubricant fraction. If necessary refining also consisted in an acid wash, agitating the oil in an "agitator" with 4-10% by weight of sulphuric acid for one or two hours, followed by a 6-8 hours settling period and a wash with caustic soda to neutralize the remaining particles of acid. Refining with chemicals was usually carried out at

---

L. P. Mongruel, Traité pratique, industriel et commerciel des huiles minérales (Paris, 1864).
A. Rey, L'huile du pétrole (Paris, 1865).
A. Dupaigne, Le pétrole (Paris, 1872).
M. Farez, Quelques traits de l'histoire du pétrole (Extr. des Mémoires Soc. agric. sci. arts de Douai, 2e série, vol. XIII, Paris, 1875).

[1] Otto Buchner, Die Mineralöle, insbesondere Photogen, Solaröl und Petroleum .... sowie die Mineralöl-Lampen .... (F. Voigt, Weimar, 1864).

temperatures of 30% and hence involved heavy losses by evaporation. The kerosine sold had a specific gravity of about 0.800-0.810.

It is not always realized that the cracking process was not introduced for the manufacture of more gasoline, but that it was applied, though in a more primitive form, at the very start of our industry. In some cases the distillation of the crude continued "until a sample of the hot residue solidified into pitch". Often the natural yield of kerosine was increased by some form of cracking. This might consist in subjecting the residue to its maximum distillation temperature for several hours during which a mild cracking took place. However, often the distillation was carried on until 5 to 15% coke was formed in the case of heavier crude and less in the case of light crude. When distilling a 0.815 spec. grav. Pennsylvanian crude the French refiners obtain [1]:

Gasoline 0.710-0.720	18.75%
Lamp oil 0.800	67.45
Heavy oil 0.850	2.83
Paraffin oil (waxy distillate)	6.63
Coke	1.60
Losses	2.74

This shows that they cracked their crude slightly and used the heavy parts of the gasoline to obtain the high kerosine yield. That this "cracking" involved difficulties in refining is clear to us now but the early refiners blamed the imperfect dephlegmation of their stills for the need for thorough chemical refining with its inherent heavy losses. Many inventors tried to circumvent it by pre-refining the crude or by distilling it with chemicals. Thus Muir [2] proposed distilling the crude with 3 to 6% of stannous chloride which had been incorporated by thorough mixing.

The solution lay indeed in the improvement of that basic operation, distillation. Up to then stills and columns had mainly been adopted from the tar and the alcohol industries [3]. Wagemann was the first to patent a process of vacuum distillation [4] and to describe its application [5]. This patent dealt with the distillation of oils obtained from bituminous shales. He blamed the bad results on varying still temperatures, the long period during which the oil was submitted to such high temperatures and excessive heating. His still consisted of two 6 ft. globes interconnected by a short cylindrical tube. The oil was

---

[1] A. Grand, Memoire sur les huiles de pétrole (Memoires et Comptes-Rendus des Travaux de la Société des Ingenieurs civils, 1869, pages 533-649).
[2] J. S. Muir, Rectification of mineral oils (Brit. Patent 3318, October 14, 1873).
[3] R. J. Forbes, Short History of the Art of Distillation (Leiden, 1950).
[4] Brit. patent Dec. 20, 1853 (Dinglers Polyt. J., vol. 135, 1854, page 138).
[5] P. Wagemann, Über die Destillation des Photogens und Paraffinöls im Vacuum (Dinglers Polyt. J., vol. 139, 1856, pags. 43-48).

desulphurized with iron-sulphate and distilled in a 26 in. vacuum, the distillates being submitted to an acid wash and neutralized with potassium hydroxide. All the "photogen" fractions were recovered below 200°C still temperature, the heavy oils below 250° and the residue contained the paraffin wax. He claimed to have reduced the losses from 8% to 2-3%.

Wilson was another pioneer of vacuum distillation [1] but he adapted the vacuum still of the sugar manufacturer, fitted it out with a steam coil and added a condenser between this still and the vacuum pump. Vacuum distillation, however, did not become popular until the higher fractions of petroleum became more important.

More successful were the first attempts at continuous distillation, made by d'Arcet [2]. He worked with a series of three stills, the residue of the first flowing into the second, etc. (Fig. 12). The stills were kept at different rising temperatures, each being heated by means of a metal bath containing a specific alloy. However, even in the case of continuous distillation, the concentration on the manufacture of kerosine often made refiners stick to their batch distillation system, though they tried to improve the fractionation of their stills. Chemical engineering was still a subject unknown and the refiners usually tried to adopt systems which had shown their merits in the alcohol industry. One of these was the Coffey still [3] "devised by Mr. John Ambrose Coffey, C. E. for the distillation of oils of almost every degree of specific gravity, ranging from the bituminous and tarry substance which is the refuse of distillation to the most rectified spirit, without intermission and in one operation". The oil to be distilled was led through a metal coil immersed in a molten zinc bath to be heated to a maximum of some 425°C and then into the rectifying column (Plate III):

"The still shown in the engraving is constructed to produce twenty products all differing in specific gravity, but the number is immaterial, as the still can be divided into any number of parts, either more or less, without in any way affecting the principle of fractional distilling. In the arrangement for obtaining twenty products, from one to six inclusive, commencing with the uppermost, are obtained highly rectified spirit; from seven to sixteen inclusive, are produced the burning oils; and from seventeen to twenty inclusive, the lubricating oils and paraffin. The residuum varies in quantity with various oils; with a moderate sample it amounts to from $2\frac{1}{2}$ to 3 per cent. The details of the

---

[1] M. G. F. Wilson, Distillation des pétroles (Le Technologiste vol., XX, 1859, page 478)

[2] J. E. d'Arcet, Distillation et rectification du goudron, de la résine, des huiles essentielles, du bitume et autre matières (Le Technologiste vol., XX, 1859, pags. 67-69).

[3] Ambrose Coffey's Patent Fractional System of Distilling (The Engineer, Nov. 23, 1866 page 364).

construction will be readily understood on reference to the engraving, where Fig. 2 represents the metal bath with the coil of pipe in it. The remaining figures illustrate the plans of the coils in the still and in the condensing cistern respectively.

When it thought advisable not to produce the most highly rectified spirit the gas may be utilised for the purpose of lighting the works, but the spirit will, doubtless, become highly valuable and be applied to new uses. This process of distilling may, doubtless, be applied to the distillation of pyroligneous acids, ethers, &c. The still above described produces about thirty-three gallons per hour continuously.

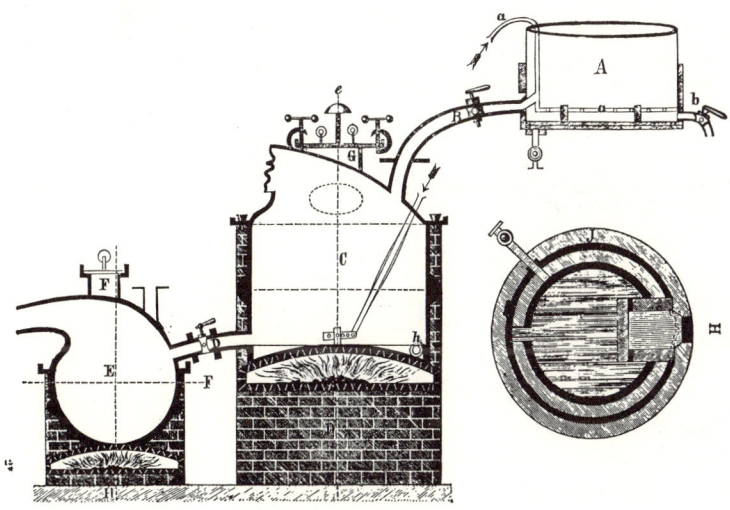

Fig. 7. Continuous distillation in two stills according to H. Perutz (1880).

We have witnessed the process in operation at the Fractional Distilling Company's Works, Millwall, and it is only justice to state that it is a complete success. The spirits and oils are beautifully clear, and the ratio in which the specific gravity of the distillations increases, from the most rectified spirit to the refuse matter, is remarkably uniform. The metal bath, which only requires a very small fire beneath it, can be kept entirely distinct from the still, and thereby preserve perfect safety. It is only necessary to keep the metal in the bath in a molten condition, and the supply cistern supplied with crude oil, and the process continues without intermission, thereby reducing the manipulation required to an unimportant item."

Another system adapted from the manufacture of alcohol was the Lugo still [1] which was patented in the United States and several European countries. Lugo distilled in partial vacuum and used a pre-heater. He tried to avoid overheating of the still bottom by not only injecting steam of 25-30 lbs/sq. inch but also air heated to a temperature 15°C above that of the liquid of the still. After three hours this temperature would be about 150°C, then chemicals were introduced to avoid the formation of ammonia. Carbon dioxide could be used if the distillate had a tendency to discoloration. In the case of "coal oil" some sal ammoniac added to the contents of the still would procure the desired "blue bloom" (Fig. 9).

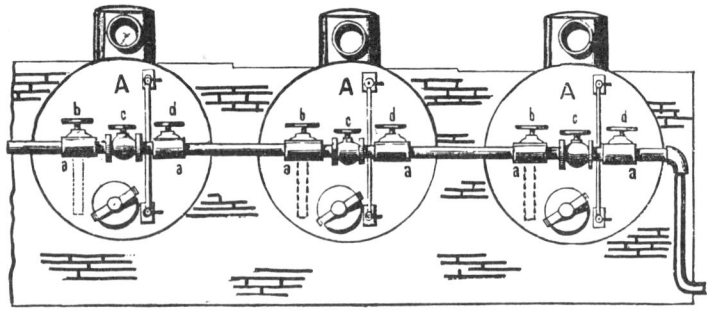

Fig. 8. Bench of stills connected for continuous distillation as used by Nobel at at Baku (1880).

One of the foremost refiners in this period was H. Perutz who wrote two books on refining. The first published in 1868 [2] deals both with the dry distillation of wood, lignite, peat, and bituminous coals (from which photogen, solar oil and paraffin wax are extracted) and with the refining of petroleum (the byproducts of which are paraffin wax, lubricating oils, greases for carts and waggons, and bituminous roofing felt). Here we find the first plan for a complete refinery (Fig. 11), in which 1 is the stillhouse for the crude, 2 the stillhouse of the wax factory, 3 the boiler house, 4 the distillate tanks, 5 the chemical refining section, 6 the wax presses, 7 the tanks for finished products and 8 those for the crude to be treated. This refinery was conceived to treat 20 Tons of crude a day by batch distillation in two large stills of 10 Tons each. By the addition of a third still sufficient reserve capacity would be installed to ensure continuous production. The redistillation of the residue in the wax

---

[1] A. Ott, Lugo's Destillirapparat für Petroleum (Dinglers Polyt. J. vol., 185, 1867. pags. 194-196).

[2] H. Perutz, Die Industrie der Mineralöle (C. Gerold, Wien, 1868, vol. I).

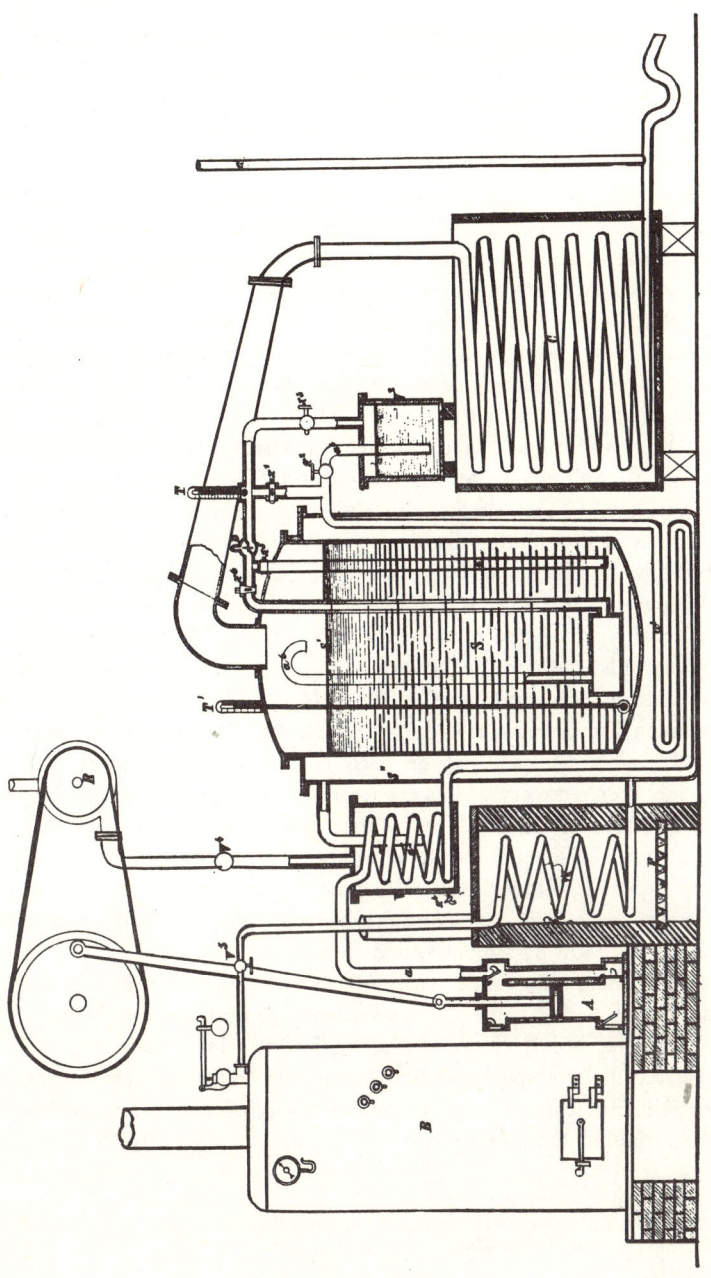

Fig. 9. Diagram of the Lugo still (1867).

factory demanded five stills of 5 Tons each. Perutz adds that the dangerous gasoline fraction can be sold with a profit to plants extracting wool or oilseeds and to carburize illuminating gas, but most of the crude oils worked in Europe were heavy and the gasoline ("petroleum naphtha") fraction caused the refiners less trouble than in other countries.

It is interesting to peruse the preface of the second volume of the Perutz handbook [1] because the author there gives a list of the major improvements in petroleum refining during the twelve years' interval between the first and second volumes. Apart from some dubious points such as "the discovery of gold in petroleum" and "the synthesis of petroleum from cast-iron and sulphuric acid" we hear of the improvements in refining paraffin wax and natural waxes, the production of vaseline and the discovery of oxygen (saponifiable materials) in certain natural waxes. Then there is the transport of oil by pipeline, the possible manufacture of aromatic hydrocarbons from petroleum and the use of light gasolines (Cymogen and Rhigolene) as an anaesthetic and for the manufacture of ice. But the main invention, according to our author, is that of continuous distillation for this meant a great saving in fuel, a better quality distillate and less cracking of the paraffin wax.

Actually, what Perutz means is not yet a true continuous distillation achieved in a series of stills (Fig. 7), but a "topping" operation which is made continuous by keeping the still level constant, continuously supplying fresh oil. By various means the residue can be tapped at intervals and distillation is stopped only now and then for cleaning out the still. The first man to operate such stills was Fuhst [2] who claimed that he doubled the performance of each still, obtained a higher recovery of less-cracked distillate, saved much fuel, reduced the wear and tear of the still and its masonry and achieved a better quality of paraffin wax. The level in the still could be kept higher and thus the vapours were less exposed to overheating. Condensation took place in condensers placed behind each still and by multiplying them one could easily replace the usual 3,500 l. still by a 6,000 l. one (Fig. 13).

In Fuhst's design the crude tank (B) is fed by pipe (A) and it serves to distribute the oil to the stills. The feed of each still is regulated by a float (F). Through the manhole (N) chemicals can be introduced into the still even during operation by a special device. Once every shift the residue is tapped from the still into the waggon (P), the contents of which are graduated to take the calculated amount of residue, specific for each crude, and sucked

---

[1] H. Perutz, Die Industrie der Mineralöle (G. Gerold, Wien, 1880, vol. II).

[2] H. Fuhst, Über die continuirliche Destillation des Petroleums, der Mineralöle, etc. bei constanten Niveau und fractionirter Condensation (Dinglers Polyt. J. vol. 207, 1873, pags. 293-304).

into the waggon by vacuum. A partial vacuum can be established by using the exhauster in the mouth of the still and fractional condensation can be obtained by the appropriate number of spiral condensers. Fuhst uses a bench of stills divided into two parts, each having its own feeding tank, and thus he is able to deal with two crudes at a time.

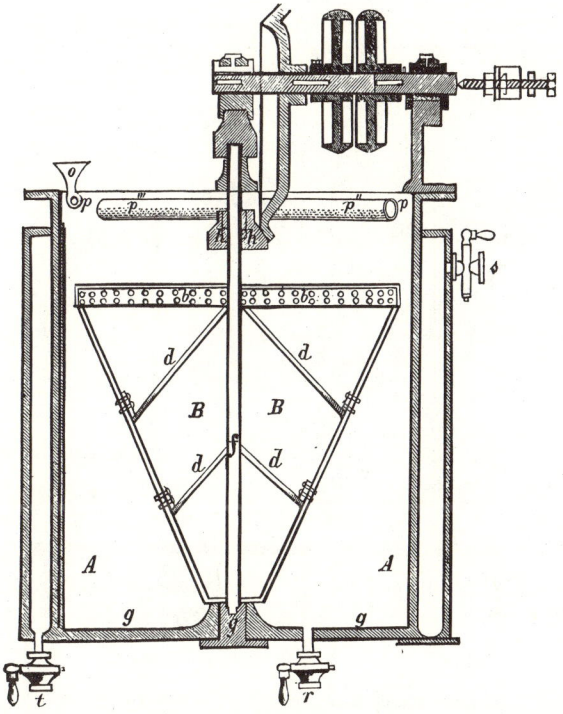

Fig. 10. Section of an early agitator and mixer.

Perutz tells us that he has improved on the Fuhst design in the new Przemysl (Galicia) factory. He has indeed established a continuous flow of oil from the feeding tank (A) to the stills (C) and (E). He uses 18 Tons stills and achieves three runs a week with waxy crude, but with other crudes he can distil continuously for at least three days without having to empty his last still. His process is halfway to that later established in Baku where a bench of seventeen stills was installed in 1880/81 to make much longer runs (Fig. 8).

Perutz uses the Coffey still for the redistillation of the light distillates though it is complicated with its 20 different condensers and flasks. He also mentions the use of compressed air for moving oil products in the refinery, in fact the "monte-jus" [1]. The bad smell of certain kerosines is due to sulphur compounds and it can sometimes be removed by agitating in vacuum, by

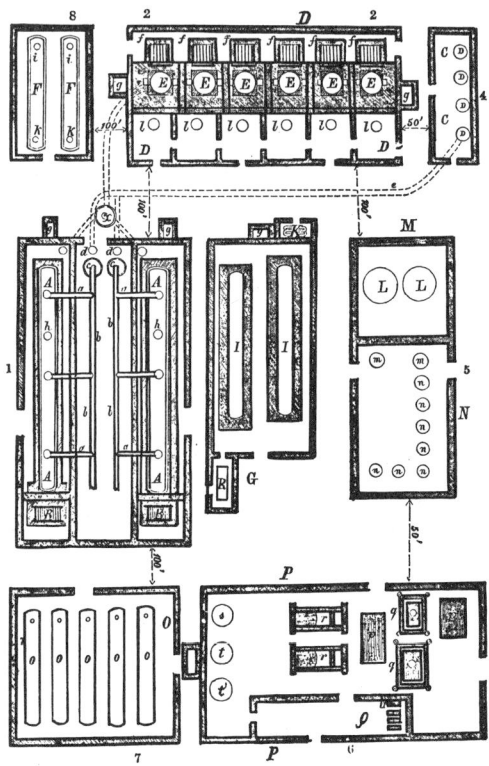

Fig. 11. H. Perutz's plan of a complete refinery lay-out.

which operation [2] the flash point rises to over 43°C. Other remedies are treating with calcium chloride [3] or with sodium plumbate [4], which latter process was to become famous (much later) for the refining of oil products containing mercaptans.

[1] L. Ramdohr, Dinglers Polyt. J. vol. 216, 1875, page 158.
[2] J. Green, Dinglers Polyt. J. vol. 180, 1866, page 144.
[3] Dinglers Polyt. J. vol. 183, 1867, page 165.
[4] Wagners Jahresbericht 1866, page 676.

OIL FOR MILLIONS OF LAMPS

The contrast with American methods is obvious when we turn from these refineries to that of a contemporary American one [1] (Plate IV):

Products of Petroleum

"The products of petroleum which can be obtained by fractional distillation, in connection with other processes employed in rectification, are very numerous; but the more important are the heavy lubricating oils and kerosene oil. A lighter product, called gasoline, is used for illumination in variosu

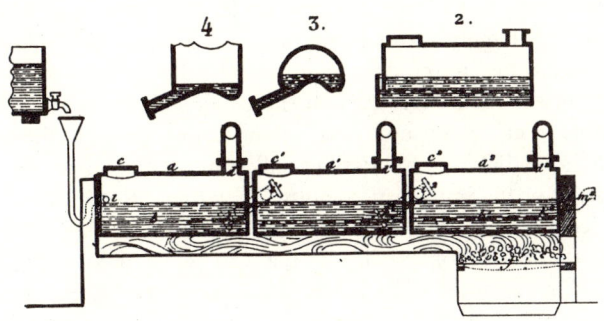

Fig. 12. D'Arcet's attempt at continuous distillation in three stills.

machines devised for the purpose. Naphtha is largely used, for the adulteration of kerosene oils, by unprincipled retailers. A product called cymogene, still lighter and more volatile than gasoline, has been utilized by Professor Van der Weyde in his ice machine. A great variety of names has been originated to characterize the products obtainable from petroleum, some of which are mentioned below, but their discussion in this place would not enlighten the general reader.

Refining Petroleum, and Separation of its Products

The crude petroleum is first put into stills and, the fires being started, a light product comes over, consisting of a mixture of substances, to which the general name of naphtha is given. The next product is called burning oil. Then follows a small quantity of paraffin oil, leaving a residuum of tar and coke in the stills. The distillates are next treated with sulphuric acid to bleach and deodorize them. The bulk of the acid is then washed out with cold water. What remains is neutralized with caustic soda, supplemented with a little ammonia.

[1] Scientific American May 18, 1872, pags. 341-343.

This brief general description of the process of oil refining gives the reader but a meagre idea of the magnitude of some of the works engaged in this business, or of the details of the operation. To give a just conception of the vastness of the petroleum industry in this country, we have prepared elaborate engravings of the celebrated

Pratt's Astral Oil Works

located in Williamsburg, N. Y., (Brooklyn, East District), which is one of the largest of its kind in the United States, and which does an enormous export business in addition to its large domestic trade, in which the article known as Astral Oil figures largely and is justly esteemed as one of the safest and best kinds of kerosene sold in the American market. This establishment is a model in its way. Every appliance that the best engineering skill could supply has been made available to ensure economy, safety and convenience, in the prosecution of the business; and although it has been alleged that the works are obnoxious to the residents of the immediate vicinity, our observations lead us to believe that this allegation has no foundation in fact. Indeed, after a thorough investigation, the Board of Health of the City of Brooklyn dismissed this complaint as unwarranted. The odors which are perceptible are not in any way injurious, nor nearly as unpleasant as those emanating from the adjacent gas works; while in point of safety, it is scarcely conceivable that, with the perfect arrangements for the prevention and extinguishing of fires, any disastrous conflagration can take place.

The oil is brought from railroad termini in large barges constructed for the purpose, each of which is fitted with a large tank to hold 1,200 barrels. When it arrives at the docks, as shown in the bird's eye view which is the first of our illustrations, it is pumped through pipes into the receiving tanks.

The Pipes

are each eight inches internal diameter, and 450 feet in length. They are laid, as shown, in an open ditch, the principal reason for which is that, in case any leak should occur, it can be at once discovered and stopped.

The Receiving Tanks

are enormous structures having a united capacity of 23,000 barrels. They are made of boiler iron and stand by themselves at a distance from other parts of the works, but are connected with the latter by pipes. In this way, the crude petroleum is kept confined and prevented from giving off its odor, and the danger of ignition is obviated.

The Pump Room

contains five large steam pumps. There is one for water, one for crude oil,

OIL FOR MILLIONS OF LAMPS   127

one for distillate and one for refined oil. The fifth is an air pump, used for agitating the oil during treatment. These pumps in their combined capacity are capable of throwing 3,000 barrels an hour. There is also another large pump on the dock for refined oil and general purposes. These pumps are arranged so as to act almost instantly as fire engines, and, through a system of piping, to flood any part of the works in which a fire may originate.

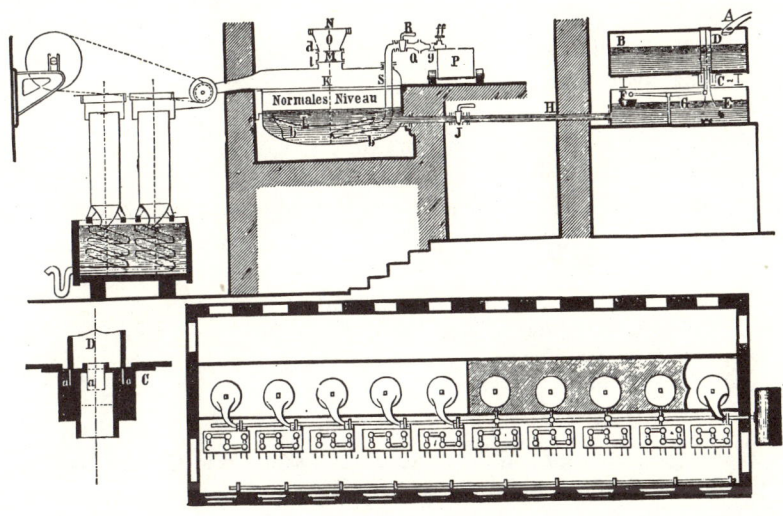

Fig. 13. Fuhst's arrangement for "continuous distillation" with continuous feed and constant still level (1875).

The office of the crude oil receivers is to separate the water from the petroleum before the latter is conveyed to

The Stills.

Four of the stills are thirty feet long and fourteen feet in diameter. Six of them are ten feet long and six feet in diameter. The stills convert the oil into vapor which passes into a condensing apparatus. This apparatus consists of a box eighty feet long, sixteen feet wide and eight feet deep, filled with coiled pipe constantly cooled by running water. The object of this distillation is to separate the volatile constituents from all solid and foreign matters. The distillation is commenced at 120°F which carries off all the products capable of evaporation at that temperature; the temperature is then steadily raised till it finally reaches about 1,000°F which drives over all except the solid residue. The distillate passes from the condensers to the

Running Room

The distillate, as it runs from the worm, appears by reflected light a beautiful French ultramarine color, but by transmitted light it has a slightly opalescent white tint. The running room is of much importance, as it is here that the progress of the work is noted, the specific gravity of the distillate is determined, and the separations are made.

There are four 500 barrel distillate tanks and two having a capacity of 1,000 barrels, to which the distillate flows from the running room. By proper connections, the stills can be connected with either of these tanks as may be necessary or convenient. From the distillate tanks, the distillate is pumped to the

Agitator.

In this apparatus, the oil is treated with one and one half per cent of sulphuric acid, by which it is bleached. It is then washed with a solution of caustic soda, followed by a little ammonia, by which the acid is neutralized; after which it is run into the

Bleaching and Settling Pans,

each of which contains 700 barrels. From the settling pans, the oil is pumped to a receiving tank having a capacity of 4,000 barrels. From this tank, it passes to the

Packing House,

placed at some distance from the other buildings and near the shipping dock. An interior view of this building has been furnished by our artist. The oil is carried into this building in pipes, where it is measured out, through an ingenious filling apparatus, into the cans which are to receive it, the cans being filled nearly as fast as they can be passed along by the workmen. The filling apparatus is shown in the centre of the background of the packing room; and, when fully employed, a dozen or more men can work at it, receiving empty cans from others as they are brought from the tin shop, and when they are filled, passing them along to other workmen who as quickly seal them hermetically with solder. They are then packed in wooden cases, ordinarily two in a case. In such packages, the cans are passed to the

Shipping Dock

At the time of our visit to the works, two large vessels lay there, taking in cargoes of oil. One of these vessels was destined to the Mediterranean, and the other was bound first to Odessa and thence to Beyrout.

The various shops connected with the works are supplied with power from two 60 horse power boilers, made at the West Point Foundery. Among these shops, we may mention first the

PLATE IV

Stills and Agitator as used in the early American Petroleum industry (After Scientific American, 1880).

PLATE V

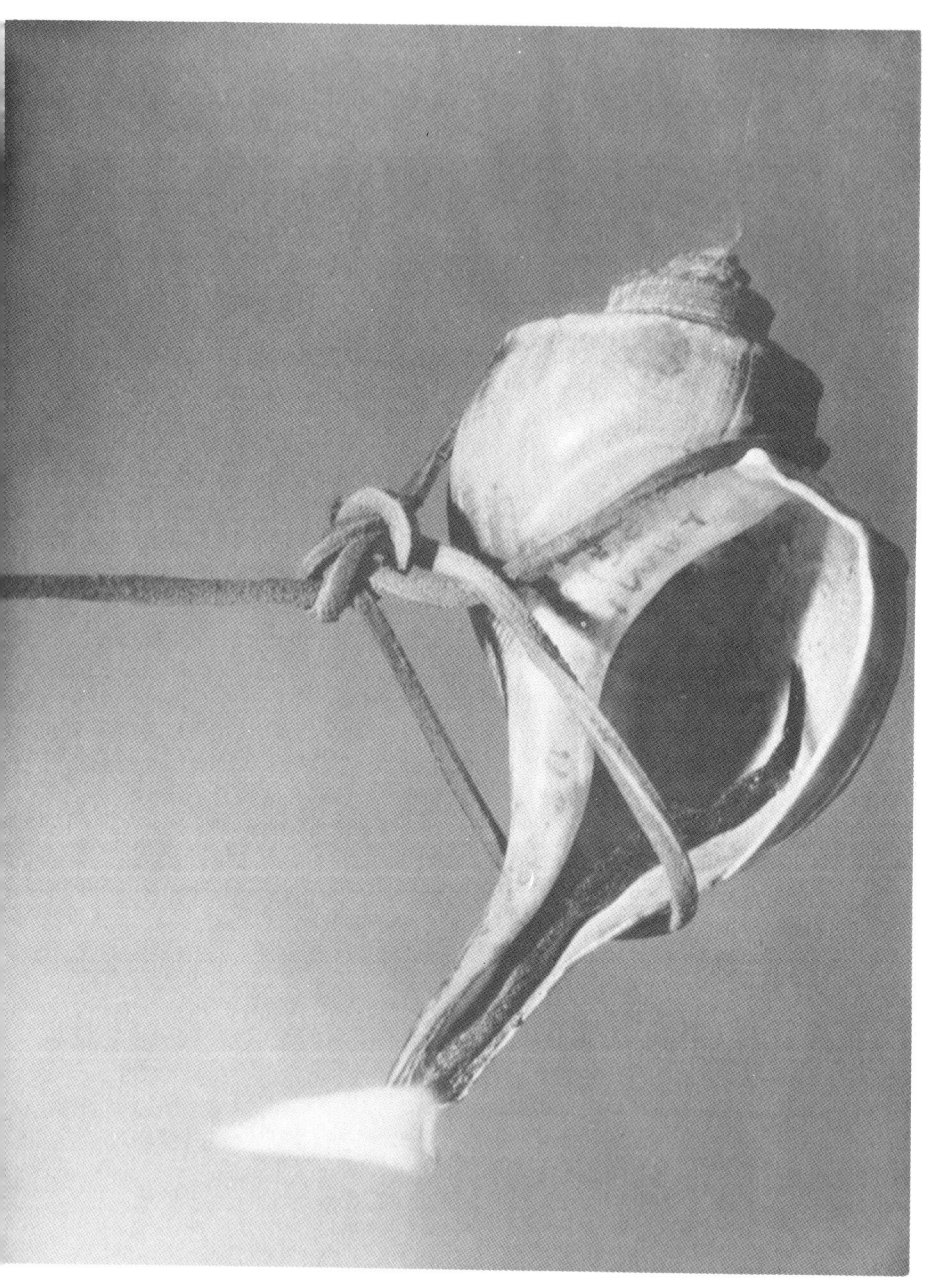

Waarschuwing.

BURGEMEESTER en WETHOUDERS van *ROTTERDAM;*

In aanmerking nemende:

dat van tijd tot tijd aanvoeren plaats hebben van uit Amerika afkomstige *petroleum*, ook wel *aard-*, *steen-* of *rotsolie* genaamd;

dat deze olie zeer ligt ontvlambaar is en vatbaar voor ontploffing, en dat alzoo bij laden, lossen, vervoeren, bewerken, opslaan of bergen van deze vloeistof bijzondere voorzorgen noodzakelijk moeten worden geacht;

herinneren bij deze aan de belanghebbenden de bepalingen van art. 16 van de verordening op de brandweer, waarbij het is verboden om dergelijke stoffen of goederen te bewaren of te bewerken, anders dan op plaatsen of in gebouwen door Burgemeester en Wethouders aangewezen of goedgekeurd, onder kennisgeving voorts dat te dezer zake aan de brandweer, de haven- en gewone politie bijzondere waakzaamheid is aanbevolen.

ROTTERDAM, den 11 Augustus 1862.

BURGEMEESTER en WETHOUDERS voornoemd,

De Secretaris,
J. L. NIERSTRASZ.

De Burgemeester,
J. F. HOFFMAN.

Gedrukt bij de Wed. P. VAN WAESBERGE en ZOON.

Fig. 14. First Dutch decree on the storage of petroleum (Rotterdam Town Council, 1862).
Studies in Early Petroleum History

Tin Shop

in which the cans are made with a rapidity truly astonishing. All the work, including the soldering, is done with improved machinery and appliances, by which the operation of can making proceeds with the utmost facility, the cans being sent as made on an endless belt directly to the packing room.

The absolute time consumed in making a can, filling it with oil, and sealing, is two minutes, the cost of the labor not exceeding two cents."

Testing the new kerosine

Both the development of the new lamps and the selection of the new lamp oils depended on the appropriate instruments to measure the luminosity of the flame, the "candle-power" of the lamp or the unit of fuel. One of the first to realize this truth was Benjamin Thompson, Count Rumford, who worked for the king of Bavaria in the early nineteenth century. He wrote to the Royal Society telling them that he was busy finding out what the most economic form of lighting was for a large workshop or public workhouse, such as he had ordered to be built in the suburbs of München. He then invented a method to measure "the relative amount of light, which lamps of different construction, candles, etc. give, which method is very simple and, I have reasons to believe, very accurate too". Such photometers showed that lamp oils should be refined, but as far as the lamp oils, then available in Europe or the U.S.A., went there was no large difference between them.

Two American scientists, Booth and Garrett, made a very thorough investigation of the lamp oils available in 1862 [1] which showed how inferior the common fatty oils were as illuminants, apart from the bad smell which could not be entirely removed. Tallow oil or distillates from resin and wood were scarce, expensive and bad. The new kerosine was proved to be the lamp oil of the future, as the following figures show:

Name of Material.	Quantity for an equal amount of light.	Price of unit of quantity.	Cost of quantity for equal light.
Gas,	1000 cub. ft.	$ 2.25 per 1000 c.F.	$ 2.25
Petroleum,	$2^5/_8$ galls.	0.45 per gall.	1.07
Candles { Spermaceti,	37 lbs.	0.50 per lb.	18.50
Candles { Paraffine,	$36\frac{1}{2}$ lbs.	0.32 per lb.	11.68
Candles { Adamantine,	47 lbs.	0.25 per lb.	11.75

[1] James C. Booth and Thos. H. Garrett, Experiments on Illumination with Mineral Oils (J. Franklin Instit. vol. XLIII, 1862, pags. 373-380).

Name of material burned	Material — how burned	Quantity burned in same time as 1000 cub. ft. of gas.	Relative cost of material burning in equal time	Cost for an average winter evening of 5 hours.
Philad. Gas,	5.1 cub. ft. per hour.	1000 cub. ft.	$ 2.10	$5^1/_3$ cents.
Petroleum	Large lamp.	$1^5/_8$ galls.	0.73	$1^7/_8$ cents.
do.,	Smaller do.	1 galls.	0.45	11-7 cents.
do.,	Chamber do.	$^1/_3$ galls.	0.15	$^3/_8$ cents.
Spermaceti	Two candles,	$7^3/_4$ lbs.	3.87	$9^7/_8$ cents.
Paraffine,	do.	7 lbs.	2.24	5.7 cents.
Adamantine,	do.	10 lbs.	2.25	$5^3/_4$ cents.

Booth and Garrett used their photometers to measure the flame of a good fish-tail burner adjusted just below the point of smoking and found little difference in quality between the oils then on the market. Important were proper care of the wick and using the lamp at the maximum height of the flame. In candles too the shape of the wick proved important. They concluded that "gaslight is fixed and can not be moved within large distances like the petroleum lamp, even at reduced gas prices it costs twice as much as petroleum". They saw very clearly that both gas and petroleum held a place of their own as illuminants:

"Although they are far less explosive than burning fluid, or any alcoholic solution of a hydrocarbon, they are by no means free from liability to explosion. The danger arises from the presence of benzole (benzine), and other volatile hydrocarbons, which have not been expelled in the process of refining, and we shall probably be subjected to this danger some time longer, in consequence of the want of skill which knows not how to remove the volatile fluids, or by reason of manufacturing cupidity, which would prefer allowing them to remain in the oil, in order to increase the quantity sold. We think there is ground for asserting that the mineral oils can be made as free from danger as spermaceti or lard oil. One great desideratum is a means of determining the freedom from danger by a simple apparatus, easily and inexpensively worked. Until the character of large dealers shall have been established for the sale of oil free from danger, we will content ourselves with this general advice to those who use the oils: never to fill a lamp at night, and not to store the oil where it can become heated.

It is not to be apprehended in the slightest degree that the oils will supersede the use of gas, especially in cities and towns, nor even in many country houses. Saving of labor, convenience, and greater safety, will cause gas still to dominate over all other sources of illumination. Besides, the astonishing cheapness of the natural mineral oils, and their richness in illuminating hydrocarbons, will probably oblige gas companies to consider the advisability of mixing mineral oil gas with coal gas. The illuminating power of any coal gas we have seen may be greatly increased with advantage, and the consequent liability to smoke may be obviated by diminishing the size or improving the form of the jet or burner."

A few years later Frankland read a most important paper[1] in which he paid much attention to James Young's coal oil and the kerosine produced from the rapidly growing amounts of petroleum. He gave the following figures:

Illuminating Equivalents, or the Quantities of different Illuminating Materials necessary to Produce the same amount of Light.

Young's Paraffin-oil	1 gallon
American rock-oil, No. 1	1.26 gallon
do. No. 2	1.30 gallon
Paraffin candles	18.6 lbs.
Sperm do.	22.9 lbs.
Wax do.	26.4 lbs.
Stearic do.	27.6 lbs.
Composite do.	29.5 lbs.
Tallow	36 lbs.

"From this table was made the following calculation of the comparative cost, from different sources, of the light of twenty spermaceti candles, each burning for ten hours, at the rate of 120 grains per hour:

	s.	d.
Wax	7	$2\frac{1}{2}$
Spermaceti	6	8
Tallow	2	8
Sperm-oil	1	10
Coal-gas	0	$4^{1}/_{4}$
Cannel-gas	0	3
Paraffin	3	10
Paraffin-oil	0	5
Rock-oil	0	$6\frac{1}{2}$

Thus, from an economical point of view, the rock-oil and the paraffin-oil approach gas much more closely than any other illuminating agent hitherto invented; while the enormous quantities in which these oils are now being

---

[1] Edward Frankland, On Artificial Illumination (Proc. R. Instit. Gr. Britain vol. IV, 1866, pags. 16-23).

produced cannot fail to make them still lower in price. They may consequently be regarded as very formidable rivals of gaslight.

The following table contains the amount of carbonic acid and heat generated per hour by various illuminating agents, each giving the light of twenty sperm candles:

	Carbonic acid.		Heat.
Tallow	10.1	cubic feet	100
Spermaceti / Wax	8.3	,,	82
Paraffin	6.7	,,	66
Coal-gas	5.0	,,	47
Cannel-gas	4.0	,,	32
Paraffin-oil / Rock-oil	3.0	,,	29

This table shows to what extent the atmosphere of rooms is deteriorated by these illuminating agents. It shows also that, from this point of view, paraffin and rock-oils stand out as the best sources of light.

The importance of such a vast amount of illuminating material so cheaply obtained can scarcely be overrated in connexion with the question of the production of artificial light. Up to the present time, the refined oil from this crude petroleum (specimens of which were exhibited) has been prevented from coming into effective competition with the original paraffin-oil, owing to the carelessness with which the former has been manufactured. There is a considerable portion of light naphtha left in this oil, which renders it capable of forming explosive mixtures in the lamps wherein it is burned. Both these American oils require to be still further freed from volatile naphtha. They would then form valuable illuminating materials."

He winds up his comparison of "gas, magneto-electric light, acetylene and mineral oil or paraffin oil" with this conclusion:

"The history of artificial illumination cannot fail to impress upon us the difficulties in the way of the application of scientific discovery to the utilities of life. How long was it after the discovery of the production of gas from coal, before a manufacturer could be found to bring it into actual operation? Thirty years ago, working in his laboratory at Blansko, Reichenbach showed us the process by which we could obtain paraffin and paraffin-oil from bituminous coal; but the discovery remained unheeded for twenty years. More than thirty years ago, Mr Faraday pointed out a source of the electric light in the permanent magnet; but we are only now beginning to use it for illuminating purposes. The brilliant little spark was long looked upon as a mere scientific curiosity, and is only now beginning to flash across the sea, guiding the mariner safely into harbour, or warning him from approaching a dangerous coast. How long will thermo-electricity have to wait before it receives a similar application? In thermo-electricity we have a direct transformation of the force of heat, which we obtain with such great economy from coal, into an electric current, and this, by further education and development, might be rendered available in the production of the electric light. Hitherto, its application in this direction has been altogether unheeded, and yet, of all sources of the power necessary for the electric light, thermo-electricity evokes this power most directly from coal. In the magneto-electric light we have the great disadvantage, that the heat of burning coal must be first transformed into mechanical power, which is made to rotate the armatures of magnets, and thus produce the necessary electric current. In this transformation of heat into mechanical power there is no less than 9-10ths of the original force in the coal absolutely lost. Hence the advantage which would result from the direct application of heat to the production of the electric current."

Frankland was perfectly right in concluding that shortly after the rise of the new petroleum industry the fatty oils would begin to disappear rapidly from the market. Only two serious competitors remained, gaslight and electricity. Gaslight was an achievement of eighteenth-century scientists such as Lebon, Minkelers and Murdoch. The latter had started using coal gas as an illuminant in 1792. In 1802 he illuminated the façade of Boulton and Watt's office when the peace of Amiens celebrations took place. In 1807 the first London street was illuminated with gas. Two years later Golden Lane Pall Mall followed and from Great Britain the use of coal gas spread to the Continent. However, using fish-tail burners did not allow the full benefit of gas to be reaped and the gas industry was involved in a hard struggle against the new kerosine lamp until Auer von Welsbach invented the gas mantle in 1883 which brought the gaslights of the towns of Europe a fair chance in surviving the competition of kerosine. However, installing gas meant a large capital expenditure and kerosine had by 1870 won its definite place as the illuminant of the poor of the towns of Europe and America, and it has remained the principal illuminant of the country districts of the entire world up to the present day.

The most dangerous competitor of kerosine was Frankland's "electromagnetic light", electricity. Its development started early in the nineteenth century, but it took several generations to gain momentum. In 1800 Davy had demonstrated how an arc-light could be used to produce a very string light. Its practical application was demonstrated by Dubosq at the Paris Exhibition of 1855. A generation later the "Jablochkoff candles" (1876) were used to illuminate streets in certain large towns, where a power-house existed to ensure regular supply of "electric juice" to feed this relatively expensive type of lamp.

The incandescent lamp did not become a success until many an inventor had spent his time on its development. Many had tried their luck following up Faraday's demonstration of a glowing filament of platinum, but an economic and cheap electric bulb was not discovered until 1878, practically simultaneously by Swan and Edison. Swan demonstrated his first model at Newcastle-on-Tyne but only in 1886 was he able to produce it on a commercial scale. Hence only by the end of the nineteenth century could gaslight and electricity compete with kerosine and paraffin-wax candles and even achieve supremacy in certain parts of the world.

However, the early petroleum industry did not yet know what the future held. It studied all types of application of kerosine including the carburation of coal gas [1]. It also discovered that different applications of kerosine demanded special lamps and specific oils. Thus when kerosine lamps were introduced in light-houses instead of vegetable oils, new specifications had to be drawn up [2]. Some of these strike us as needless, such as the French instructions which demanded a coefficient of dilatation of the oil of 0.00074, whereas scientists had already established this figure to be 0.00094 for coal oils and kerosine. It seems that the French lighthouses mostly used "Scotch paraffin" supplied by Young's Paraffin Light and Mineral Oil Company. Gullo tells us "that (like kerosine) it emits a slight odour of rock oil which is not disagreeable. When persons for the first time inhale the emanations from paraffin they suffer slightly from headache, but this symptom disappears as they become accustomed to the vapour". The average analysis of such lighthouse kerosines, as established in the laboratories of the "Ecole des Ponts et Chaussées" and the "Dépôt des Phares" during 1869 was:

Intensity of illuminating power corresponding to the consumption of 40 grammes per hour	2.18
Temperature of ignition, or grade of inflammability	72°cent
Boiling temperature	205°cent
Density at 0°	0.833
Co-efficient of dilatation from 0° to 100° cent.	0.094

The report proceeds to describe the vessels and the magazines used for storing the oil. A second portion of the paper treats of the lamps used for

---

[1] B. H. Paul, Artificial Light (J. Soc. Arts, vol. XII, 1864, page 311 ff.).
B. H. Paul, Carburation of gas (J. Soc. Arts, vol. XI, 1861, pags. 503, 520). C. A. Martius, Illuminating gas (Amer. J. Pharmacy (3) vol. XVIII, 1870, pags. 326 ff.).
E. Collin, Eclairage aux huiles minérales (Paris, 1870).
[2] L. Gullo, Sull'Applicazione del Olio Minerale all'Illuminazione dei Fari (Giornale del Genio civile, vol. XV, 1874, pags, I, 61, 121, 421).

consumption of mineral oil, which are not substantially different from those used for that of vegetable origin. The principles of the lamps in use are these: they are either lamps with reservoirs below, which are fed by capillary attraction; lamps in which the oil is kept at a constant level, by a raised reservoir; or lamps with an inferior reservoir, but regulated supply, as in the case of a moderator lamp. The Maris lamp, one of the first group, was used in the French lighthouses in 1857. The diameter of the burner was 1.45 inch, the diameter of the interior current 0.78 inch, and the height from the corona of the burner to the top of the reservoir 1.3 inch. Since the year 1868 the introduction of the method invented by the American, Captain Doty, which maintains a supply of oil at a constant level to a lamp with four wicks, had given rise to law-suits yet pending. Minute details are given of the Doty system, which is compared with that of Fresnel. The hourly consumption of ten different lighthouses is tabulated, the mean being 950 grammes (33.5 ounces avoirdupois). The report concludes with a list of the principal French manufacturers of lamps used for the lighthouses of France.

The early refiners were rather worried about the sulphurous smell of their kerosines. Thus Vohl [1] mentions that all kerosines contained some sulphur which might endanger the health of the consumer. As the only crudes then treated came from Pennsylvania, Canada, Galicia, Wallachia (Roumania), Caucasia or Rangoon, the fault could hardly lie in the crude itself but was probably due to inefficient chemical refining methods, which were soon remedied. However Vohl showed that none of his 28 samples were sulphur-free and as they also contained appreciable traces of acid, the cause was probably an insufficient removal of the acid tar in the refineries. Vohl insists that the refiners should direct their attention to this problem, for the production of kerosine was rapidly expanding. It had proved to be seven times less expensive than gaslight. The bluish bloom which some kerosines possessed was due to traces of higher boiling compounds.

The greatest worry of the early refiner lay, however, in another direction. In their desire to produce as much kerosine as possible from their crude, they incorporated large portions of heavy gasoline in the true kerosine fractions. This made their kerosine highly inflammable and dangerous to handle and to store. Vohl states that the presence of such "naphtha" is easily detected by the flash point of the kerosine which should not be lower than 27.5 to 28.5°C. White of New Orleans had pointed out that the average American kerosine was heavily adulterated with naphtha, which lowered its

---

[1] H. Vohl, Über das Petroleum als Beleuchtungsmaterial, seine Verunreinigungen und die durch letztere veranlasste Entwicklung gesundheitsschädlicher Gase während des Verbrennens (Dinglers Polyt. J. vol. 216, 1875, pags. 47-51).

flash point to 23°C and even less, meaning that up to 10% of naphtha had been added. He stated that such a compounded oil burns considerably worse as the adulteration with naphtha is accompanied by the addition of heavy fractions which do not belong in the kerosine either. The latter gradually char the wick considerably and thus cause a rapid fall in luminosity of the lamp.

Coleman too [1] was worried about this adulteration with naphtha and heavy fractions. Whereas the European paraffin refiners had steadfastly adhered to the principle of separating all volatile hydrocarbons so carefully from burning oils intended for general use, that it would be difficult to find any sample of paraffin burning oil in the market that would give off inflammable vapour under 120°F, the bulk of American petroleum-burning oils were so far contaminated with naphthas that it was very difficult to get a sample of imported petroleum which would not give off inflammable vapours at from 90° to 100° F, tested in the same manner as would result in a flash point of 120° for the paraffin oils. The habit adopted by the Scotch refiners in sending out safe oil was due to a regard to the public safety. When crude oil of either kind was deprived of its tarry, basic, and acid impurities by successive treatment with acid and alkali, the purified hydrocarbons were fractionated into groups, such as naphtha, burning oil, lubricating oil, &c., by the process of fractional distillation. By repeated distillations American petroleum is often broken or "cracked" up into

Crude naphtha	20 parts
Burning oil	66 ,,
Coke and loss	14 ,,
	100 ,,

True, the American refiners used some of the lighter portions of their crude for "air gas" and burned vast quantities of crude naphtha for heating purposes, but they "put as much as they dare in the burning oil; they are only stopped from going to an unlimited extent by the legislative acts of the foreign countries to which the material is exported". No other market was then available in the United States.

Of course tests were soon developed to check the addition of naphtha, the distillation test and the determination of the flash point. Several attempts had already been made by the chemists to fractionate petroleum. Neither Eisenstück [2] who distilled crude oil from Schude (Hannover) nor Peball [3],

---

[1] J. J. Coleman, On the Methods in Use for Testing Hydrocarbon Illuminating Oils (Chemical News, April 2, 1875, pags. 147-148).
[2] Ann. Chemie Pharmazie vol. CXIII, page 169.
[3] Ann. Chemie Pharmazie vol. CXV, page 20.

working with a Galician crude, could obtain constant boilsing fractions from the original 5° fractions they distilled from the crude. Warren [1], however, distilled crude oil into fractions, measuring carefully their distillate for every 20°C and submitting each of these fractions to frequent redistillations. He claimed to have obtained "these constituents so pure, that the contents of an ordinary tubulated retort charged with one of them completely distilled off without essential change of temperature (not the amount of $\frac{1}{2}$°C). This state of purity, I think I may safely assert, has never before been attained from such mixtures by any system of fractional distillation.

It is not too much to anticipate that, whenever the various constituents of the mixtures referred to shall have been separately and thoroughly studied in a pure state, some of them may be found to possess properties which will give to them great commercial value, sufficient to justify the expenditure necessary to separate them in large quantities".

Such fractional distillations of samples led to the wellknown distillation test. Letheby [2], in imitation of a similar test for coal-tar and coal-tar fractions, proposed an apparatus for the analysis of "the essences of petroleum oils for commercial purposes". He used an ordinary distillation flask and a Liebig condenser and observed the temperature at which the first drop of the liquid falls into the receiver, weighing the total distillate collected as soon as the required temperature of the liquid in his flask has been attained. From such apparatus the well-known Engler, I. P. and A.S.T.M. distillation tests were derived.

Coleman assures us that adulteration with naphtha is "detectable by fractional distillation test as devised by Mr. Valentin in 1871. The Board of Health Dept. of New York and Franklin Institute in 1869 tested no less than 639 samples of which only 21 did not give inflammable vapour at 100°F, in 1871 out of 100 samples only 7 were safe oils". It is more difficult to establish a good flash- and fire-test. In the opinion of Coleman Tatlock's apparatus was considered best, as the rate of heating was properly defined. "It is most scientific but brokers and commercial people will scarcely accept it". Other tests gave widely varying results "differing to the extent of 10 to 15°F", and this was the weak point in the British Petroleum Act of 1871 until the Abel apparatus [3] was accepted.

---

[1] C. M. Warren, On a process of Fractional Condensation; applicable to the Separation of bodies having small Differences between their boiling points. (Mem. Amer. Acad. Arts and Sciences vol. IX, 1867, pags. 121-134).

[2] Dr. Letheby, Apparatus for the fractional distillation of volatile oils of petroleum, coal-tar, etc. (J. of Gas Lighting, Water Supply & Sanitary Improvement 1863, pags. 653-654).

[3] F. A. Abel, Report to the Secretary of State from the Home Department on the subject of testing petroleum (London, 1877).

In the United States local acts specified a fire test of 100 to 120°F and a flash test of 100 to 110°F for commercial kerosines, but in European countries conditions were more stringent and special care was taken to specify the conditions under which these inflammable goods could be stored. Thus in France, as Rey [1] informs us, the "naphthometer" is used to determine the flash-point, which should be below 51 to 55°C but above 35°C. Kerosine should be stored in a properly ventilated and cool space. No lots of over 150 litres should be kept within 45 M. distance of houses or public buildings. The "Conseil d'hygiène et de Salubrité publique de la Seine" had drawn up an instruction for the general public how to handle lamp oil and oil lamps which instruction was approved by the police commissioner on June 29, 1864. The conditions for the storage of petroleum were then laid down in the decree of April 18, 1866, which was turned into a law on December 31, 1866. It specified a minimum flash-point of 35°C. As facilities for storing drums of petroleum were being built at Marseilles a special decree for their safe storage was given in a circular of July 15, 1868 [2].

In Great Britain there were a series of Acts from 1862 onwards laying down the conditions under which this inflammable petroleum was to be stored [3]. Petroleum was defined in these words:
"For the purposes of this Act (1871) the term "petroleum" includes any rock oil, Rangoon oil, Burmah oil, oil made from petroleum, coal, schist, peat or other bituminous substance, and any products of petroleum, of any of the above-mentioned oils; and the term "petroleum to which this Act applies" means such of the petroleum so defined as, when tested in manner set forth in Schedule One of this Act, gives off an inflammable vapour at a temperature of less than one hundred degrees of Fahrenheit's thermometer, and a minimum flash-point of 73°F or 22.7°C was first allowed. The Act of 1879 incorporated a description of the Abel flash-point test.

In Holland the first ship carrying petroleum arrived in the port of Rotterdam on June 30, 1862. The City Council issued a pamphlet on August 11, 1862 warning dealers and consumers of the dangers of storing large quantities of petroleum. Certain rules for the storage of petroleum were given in a bye-law of

[1] A. Rey, L'Huile de Pétrole (Pfefer et Puky, Genève, 1865 pp. 68-69).
[2] Bizard et Labarre, Mode d'emmagasinage des huiles de pétrole, de schiste, etc. (Ann. des Mines, Mémoires 1867, pags. 185-206; Partie administrative 1868, pags. 275-276).
[3] Act for the Safe-Keeping of Petroleum, 29th July, 1862: Act for the Amendment of the Law with respect to the Carriage and Deposit of dangerous Goods, 6th August, 1866.
Act to amend the Act 25th and 26th Victoria, Chapter 66, for the safe keeping of Petroleum, 13th July, 1868.
Act for the safe keeping of Petroleum and other substances of a like nature, 21st August, 1871.
Act to continue and amend the Petroleum Act 1871, 11th August 1879.

1867 [1], which, however, does not specify any flash-point or other property of kerosine.

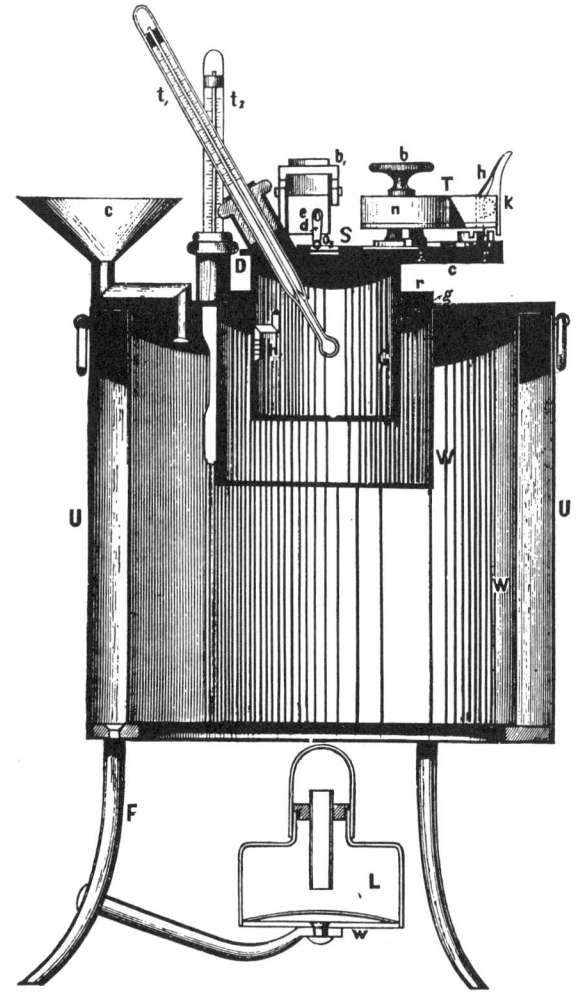

Fig. 15. Abel flashpoint apparatus as laid down in the British law of 1880.

Thus the European refiners had no difficulty in producing a good kerosine for commercial purposes during the early phase of the young petroleum

[1] Verordening op den aanvoer, het bewaren en vervoeren van Petroleum, Terpentijnolie en Benzine (Naphta) te Rotterdam (17 mei 1867).

industry. Their main worry was the development of good and economic manufacturing and refining processes, as European economy demanded that the utmost profit should be reaped from all byproducts of kerosine manufacture, such as lubricants, asphaltic bitumen and wax. The creation of such integrated refineries was their main occupation during the latter part of the nineteenth century.

In conclusion we would like to quote a few passages on the new lamp oil, kerosine, taken from America's oldest handbook on oil:

"As an illuminator the oil is without a figure: It is the light of the age. In the opinion of some who have considered the subject, illuminating is its grand office. Those that have not seen it burn, may rest assured its light is no moonshine; but something nearer the clear, strong, brilliant light of day, to which darkness is no party. It tries the eyes of none. For the Christian by means of it to peruse his Bible, is no infliction. It never causes the politician to weep, when he reads at night in his favorite newspaper, the victories of his own party; nor the merchant to shed tears over the price current, showing a turn in trade which puts money into his pocket. In other words, rock oil emits a dainty light; the brightest and yet the cheapest in the world; a light fit for Kings and Royalists, and not unsuitable for Republicans and Democrats. It is a light withal, for ladies who are ladies indeed, and so are neither afraid nor ashamed to sew or read in the evening. An oil man, without any risk of a breach of promise, may warrant them, that by this light, they can thread their needles the first time, and every time they try.

It appears nature first tried her apprentice hand and gave us fish, lard and coal oils for illuminators. Now as the product of her better skill, she furnishes us with Rock oil. If we cannot infer the fact from the theory, we can wheel about and prove the theory from the fact. For in power to give a clear and intense light, it surpasses the best sperm. When properly rectified it produces a larger, brighter flame, of uniform size. No perceptible smoke attends the combustion, and no unpleasant odor escapes into the room. When a person reads, writes or works by its aid, the work and letters printed or written are so well illuminated as to be seen distinctly, and with great ease by all sitting round a large table. Of course, the crude oil must be properly rectified and burnt with a suitable wick.

Let us compare this new-born light with some of its older sisters. Comparisons are said to be invidious; but their serving to give us a more distinct view of objects justifies their introduction in this connection.

Rock oil possesses several advantages over Camphene or Burning Fluid. The same quantity by measure burns more than twice as long. As a small compensation, the light of the Camphene is one-fifth brighter. This one-fifth in most cases is unnecessary for the light from the oil is brilliant enough. And if required, it is dearly bought; that is, assuming the price of each to be nearly the same; and it is certain the oil can be furnished at as low a figure as the Camphene or Fluid. Evidently, then, the greater economy of the oil, furnishes a strong recommendation in its favor, when compared with the other.

Again, it gives forth a more uniform light, continuing it without any perceptible decrease for 12 hours. It also burns longer before the wick becomes clogged or crusted over. This frequent crusting, when camphene is used, is a marked inconvenience. Can Kerosine be manufactured, as good authority decides, for 25 cents per gallon, to say nothing of freight, commission, &c.? Rock oil can be raised from many wells for 1 cent a gallon! At none of the principal wells does it cost in the vat 2 cents; and it can be refined for 5 cents the gallon. Add to this the fact that, 80 parts out of a 100 of the Venango oil is a prime illuminating fluid, and the balance worth as much as the whole price of refining; and is it any longer a question which is destined to be the light, providing rock oil continues abundant, as we have every reason to believe it will be? Not less sagacious than candid was the remark of a coal oil man who some time since visited Titusville: "If this business succeeds, mine is ruined".

To so clarify rock oil that it can be burned in a common lamp with a large blaze and yet without smoke, has been hitherto a desideratum. A lamp with a chimney cannot so well be carried about the house. This, however, can be done with the light reduced one-half, which will then be equal to that of two candles. But in some families other lamps are kept ready for use in going from one room to another. Such might have for their purpose a rock oil lamp without a chimney, in which a small blaze can be kept up unattended with smoke.

Here opens a broad and lovely field for the exercise of ingenuity. It will be hailed as a grand achievement when this oil is prepared for the hand lamps. We have feared it would not soon be done. But a gentleman informs us that a company called the Philadelphia Diamond Oil Co., profess so entirely to deodorize and rectify it, as to insure the desired result. If they so advertise, as he states, we are bound to credit their word, till its falsity appears. The Philadelphians are not accustomed to do things at halves; and if they have really succeeded in preparing from the raw material an oil which burns in a common lamp without smoke, they will prosecute the business on no limited scale, and many will soon receive the benefit thereof.

Before we close this chapter let us pause to congratulate each other on having been put in possession of so excellent a light; one within the reach of all, and at the same time fully a match for the city gas-blaze. Those who cannot afford to buy sperm or lard oil or Kerosine, can now enjoy an illumination superior to that of either.

In some localities, candles are used in most country dwellings. They are always running down, and when carried from one room to another are giving off drops of grease, greatly to the annoyance of the tidy house-wife. In summer, the temperature being about that at which they melt, they are a trouble indeed; and cannot be made unless it be at night. Besides the care of keeping them snuffed is a constant trouble.

Farmers whose tallow is a home production, where they cannot sell it for any price, will continue to use candles. Probably, however, it will all be wanted for other purposes soon. Many of them will not deny their eyes and the brighter ones in their family circles, for the sake of saving a few pence, the luxury of a good light, when the toils of the day are ended and all gather around the center table.

Especially do we congratulate the people of Western Pennsylvania, in view of this gift of God to them.—We have been here shut out from the world in a measure, our development checked by spurious land titles of an early day, as well as by the high price of the wild acres. We have remained nearly stationary on the bank, while the tide has rolled west, and spread itself over lower priced and more fertile soil. But in compensation for privation and poverty, our Kind Father in Heaven has caused the rock to pour us out rivers of oil and thus given us at once magnificent light for our dwellings and a source of honest wealth. The influence direct and indirect of this benefaction on our social order, will be happy, powerful and lasting."

CHAPTER SEVEN

WAX FOR OUR CANDLES

Paraffin wax, like kerosine, was destined to take the place of fats and oils, which could easily be made fit for human comsumption and thus help to fill the world's growing need for fats. Its story is intimately bound up with that of coal oil and kerosine.

Prehistory of the candle and its manufacture

The candle was probably an invention of the Etruscans [1] and the Romans brought it to Greece, where only tapers had been used before, to the Near East and to Western Europe. The oldest candles were handmade and consisted of a wick made of oakum, papyrus, rushes or linen (later also cotton) which was coated with a layer of "fuel" by immersion in molten tallow or beeswax. In the eighteenth century, and in some country districts even later, rushlights were manufactured by peeling rushes until a wick of pith remained held together by a strip of peel, which wick was then immersed in tallow to manufacture a cheap form of candle. In those days a candle cost a halfpenny and burned for some two hours; for the same amount of money one could buy twelve rushlights, which gave for many more hours a much weaker light.

The use of the candle received a strong impetus when the young Christian church adopted its use in the church ritual in late-Roman times. Thus we read that during the reign of Constantine the Great in the fourth century the town of Byzantium was illuminated with oil lamps and candles on Easter-eve. In those days the Church specified that the candles to be used on the main altar should contain at least 65% of beeswax; the cheaper tallow candles were used at home.

Both tallow and beeswax candles were made by hanging the wicks vertically and pouring the molten wax down along them until the desired size had been reached. Then they were rolled on a stone or marble slab and cut to size. Naturally such hand-made candles were relatively expensive. Later Chinese

---

[1] R. J. Forbes, Studies in Ancient Technology vol. VI (Leiden, 1958, Chap. III).

Plate VI

Typical examples of crudely made lamps still in use in many undeveloped areas
(A Shell Photograph).

PLATE VII

Candle-making at the Belmont Works of Messrs. Price's Patent Candle Company Limited, London (Shell Photograph, 1955).

wax and other vegetable waxes were also used for the manufacture of candles and when whaling grew to be an industry, spermaceti was added to the candles to give them a better appearance. Tallow candles were not popular in the South, as they soon became soft in a hot climate, but they were very popular in Western Europe.

Candle manufacture remained unchanged until the early nineteenth century. Three factors contributed to this change, the casting of candles, better wicks and new, cheaper ingredients. Casting candles in moulds was invented in the fifteenth century by a certain Sieur de Brez according to tradition, but this was certainly only a first attempt which did not become popular. Two problems remained unsolved: the rapid cooling of the filled moulds and detaching the consolidated mass from the mould.

In 1801 Thomas Binns of Marylebone (London) invented a casting press which consisted of a series of cylinders in which the wick was stretched, surrounded by a body through which cold or hot water could be circulated. This enabled the manufacturer to cool the mass at will and to obtain beautiful glossy candles. A series of inventors such as Joseph Morgan, Joseph Tuck, William Palmer, John Stainthorp and Willis Humiston helped to perfect this machine until by about 1855 automatically discharging and readjusting machinery was available.

In the meantime better wicks had become available. In the old days much time and material was wasted because the candles had to be snuffed from time to time. Composition and treatment of the wick was not yet such that it bent and burnt as the candle was consumed. Soon after lighting the candle it would char into a bulgy mass which had to be cut off with a special pair of scissors. This was very annoying for those who had to work or study by candlelight and one can understand that Goethe asked his fellow-scientists to spend their time "on a source of light which one need not snuff continuously". In 1820 the Frenchman Cambacères found the solution by using a properly woven cotton wick instead of the usual strand of cotton threads only loosely intertwined. This eliminated snuffing and later the impregnation of the wick with certain salts promoted its combustion when the bent top of the wick came into contact with the outer mantle of the flame. By 1835 Thomas Ripley had begun to manufacture such wicks in Great Britain.

Even in the days of Napoleon no better fuels were available for the manufacture of candles than the expensive beeswax and tallow. However, the quality of tallow was very variable and its melting point was far from constant. Sometimes it contained a large percentage of liquid oils, which had to be removed by expressing. In 1799 William Bolts had obtained a patent for such a procedure to separate the solid parts from tallow. Tallow candles burnt well with an oakum wick, but they had the disadvantage that they could be eaten. This disadvantage was conspicuous in lighthouses, where candles were often substituted for the oil lamps which had a tendency to smoke. When the control of the lighthouses of Great Britain was laid in the hands of Trinity House (1807) it appeared that the watchmen of the famous Eddystone lighthouse often used their tallow candles to increase their meagre rations of fat. Such stories are also told of the Russian army which helped the Allies to destroy the power of Napoleon.

Soon a much better ingredient for candles was discovered. Between 1813 and 1823 the French chemist Chevreul conducted his famous researches on the composition of oils and fats. He found that they consisted of glycerine and fatty acids. The fatty acids isolated with caustic soda and potash consisted of a liquid part, "oleine", and solid acids forming the "stearin". This stearin, the solid higher fatty acids, was an excellent ingredient for candle manufacture. A few years after Chevreul published the results of his work (1823) Gay Lussac and he took out a patent for the manufacture of stearin candles. Their ingredients were still too expensive to make them a commercial success, but shortly afterwards Milly was able to use the cheap lime solution instead of the caustic soda and by 1832 the pilot plant began to yield profits. In 1833 no less than 25 tons of stearin candles were sold in Paris at two shillings a pound, which was very cheap in those days.

In 1835 the first stearin candles were marketed in England by Messrs Edward Price and Company of Battersea, a firm founded in 1830 by William Wilson and Benjamin Lancaster, and the ancestor [1] of the later Price's Patent Candle Company Ltd., which celebrated its centenary in 1947. Their stearin candles were a great success when, during the marriage festivities of Queen Victoria and Prince Albert in 1840, illuminants of all kinds were in great demand. They were rapidly absorbed by the public and when a few years later renovation of the House of Commons was discussed and somebody proposed the introduction of gaslight, the answer was: Do what you like to improve the acoustics and the ventilation, but take it for granted that the candles remain".

However, the separation of oils and fats into glycerine and acids was still far from perfect. In 1842 Wilson and Jones patented a process using sulphuric acid, in 1855 Milly suggested the use of lime in autoclaves at a pressure of $1\frac{1}{2}$-2 atmospheres. Slowly by about 1870 this autoclave process became a commercial proposition and in 1898 Twitchell discovered the reagent which was to bear his name and which meant a great saving on the amount of sul-

---

[1] Price's Patent Candle Company, Still the candle burns (London, 1947).

phuric acid used in the autoclave process. Still the acids thus manufactured were usually darkened by the use of sulphuric acid. Refining by distillation had to follow for it was essential to make white candles. In 1840 Gwynne had patented a vacuum distillation of crude stearin in a silver still. He had joined Messrs. Price and continued his researches. It soon became evident that no vacuum distillation was needed but that the use of super-heated steam would achieve the same effect and yield beautifully white stearin. The colour of the oleine was also improved.

When the stearin candles proved such a success the search for new base materials started, for the amount of fatty oils was already limited and the manufacture of candles was using them at the expense of human consumption. One could not continue to transform food into light indefinitely. In 1829 James Soames had succeeded in separating stearin from palmoil and when Hempel and Blundell perfected this process the firm of Price decided to buy this oil on a large scale in Ceylon and Africa. For their steam-distillation process enabled them to transform these oils and even waste fats into good white stearin. But their supply was not inexhaustible and the demand for stearin candles increased.

The new ingredient, paraffin wax.

Here the petroleum industry stepped in. In the same way as its kerosine was to save valuable oils for human consumption and its modern product, synthetic glycerine, was again freeing fats for their natural destination, it was soon after its rise to produce paraffin wax as an excellent substitute for stearin. Very soon candles were to consist mainly of paraffin wax with a small quantity of stearin to achieve the desired appearance of the candle.

The discoverer of paraffin wax was Carl Reichenbach of Blansko (Moravia) where he directed a series of mines, iron foundries, shops and chemical plants on the grounds of the Count of Salm. Later he became Carl Freiherr von Reichenbach. He liked experimenting and became interested in the products of dry distillation which so many of his contemporaries were studying. Studying these tars he found that the heavy fractions showed progressive viscosity and at a certain point began to solidify and separated into flakes. Microscopic examination showed their crystalline form. Cleaned with filter-paper they proved to be of a very light colour and elimination of the oil with alcohol yielded a pure white product with a fatty appearance and touch, which was tasteless. Its melting point was about 44° C, its specific gravity 0.870. Reichenbach identified it with the "camphoride" or "stearopten" which Berzelius had described. As it could easily be recrystallized with alcohol or ether and showed no affinity to any chemical reagent Reichenbach suggested the name "paraffin" (from parum and affinis) [1]. He had already found that it mixed well with stearin and beeswax. He therefore suggested that this product of the dry distillation of wood would be an excellent material for burning

Fig. 16. Snuffing the candle (After Jan Luyken, about 1700).

instead of the vegetable or animal oils which had such a strong tendency to soot. It might also prove to be a good lubricant.

---

[1] C. Reichenbach, Beiträge zur näheren Kenntnis der trockenen Destillation organischer Körper (Jahrb. Chemie & Physik, vol. XXIX, 1830, pags. 436-460).

Gay Lussac [1] who obtained a sample from Reichenbach concluded that "it consisted entirely of hydrogen and carbon in a ratio 2 : 1 just like olefiant gas (ethylene)".

The products of dry distillation were not easy to analyse and certainly not in the days of Reichenbach. Laurent analysing the bituminous Autun shale "from which Messrs. Blum make their illuminating oil" found that it contained 53% combustible material on dry distillation [2]. Cooling the oil down to $-5°$ C. white scales are deposited which can be recovered and cleaned with alcohol, They proved identical with the paraffin wax obtained by dry distillation of coals, shales and wood. As these scales can not be obtained by extraction of the shale they must have been formed during the distillation.

Reichenbach continued his work and, wanting to prove Dumas' contention that "naphtha" is a product of dry distillation and probably a mixture of polymers of olefiant gas like paraffin wax formed during this process [3], he tried to isolate individual hydrocarbons from the lower boiling fractions. One of these he called Eupion. His theory engendered a long paper war between him and Hess from St. Petersburg [4] which is no longer of much interest to us, as neither knew much about the composition of their distillates.

Laurent confirmed Reichenbach's ideas [5] and stated that the lower boiling fractions of the shale naphtha, like the paraffin wax, seemed to belong to a series of hydrocarbons with the formula $C_n H_{2n}$. He also separated from the heavy fractions a substance which he called "ampeline" "which resembles creosote". Lewy [6], using the new atomic weight of carbon proposed by Dumas and Stass, was able to correct the view that paraffin wax was a polymer of olefiant gas. Its formula was found to be $C_n H_{2n+2}$ but his analytical tools were not sufficiently accurate to determine, whether the formula should be $C_{20}H_{42}$ or $C_{48}H_{100}$.

The mineral wax Scheererite was discovered in a bed of lignite (brown coal) at Uznach, near St. Gall in Switzerland, by Captain Scheerer, in 1823. In the

[1] J. Gay-Lussac, Analyse de la paraffine (Ann. Chim. Phys vol. XXXXIX, 1832, pags. 78-80) (Ann. d. Physik, vol. XXIV, 1832, pags. 173 & 180).
[2] Aug. Laurent, Sur les schistes bitumineux et sur la paraffine (Ann. Chimie Physique vol. LIII, 1833, pags. 392-396).
[3] Dumas, Ann. Chim. Phys. vol. L, 1832, page 238.
[4] C. Reichenbach (Scheigger's Jahrb. Chemie Physik vol. LXII, 1831, page 150 ff.); H. Hess, Über einige Produkte der trockenen Destillation (Ann. Physik, Chemie vol. 36, 1835, pags. 417-436).
C. Reichenbach, Über Eupion und Bergnaphtha in Bezug auf die Ansichten des Herrn H. Hess (Ann. Physik Chemie vol. 37, 1836, pags 534-544).
[5] Aug. Laurent, Sur l'Huile des Schistes bitumineux, l'Eupion, l'Acide Ampélique et l'Ampéline (Ann. Chim. Phys. vol. LXIV, 1837, pags. 321-328).
[6] M. Lewy, Note sur la composition de la paraffine (Ann. Chim. Phys. vol. V, 1842, pags. 395-399).

same year the mineral wax hatchettite or hatchetine was discovered on the borders of Loch Fyne, in Argyllshire, Scotland, and was named after the English chemist, Charles Hatchett. The first reference to the mineral ozokerite was by E. F. Glocker in 1833. He discovered it near the town of Slanik in Moldavia, close to a deposit of lignite at the foot of the Carpathians. It was named from the Greek words signifying "to smell" and "wax", in allusion to its odor.

In 1833 Meyer had again drawn attention to the natural waxes already known to the geologists and he had proved that ozokerite consisted of a mixture of waxes which could be distinguished by their solubility in alcohol, their melting point and specific gravity. Moreover it was realized that many products of dry distillation of various materials and also crudes contained paraffin wax. Wagemann [1] published a list of such oils containing from $\frac{1}{2}\%$ to 3% of paraffin wax.

From 1850 onwards James Young was manufacturing paraffin wax from his Boghead Coal at Bathgate and Reichenbach [2] was the first to acknowledge that though Young had not discovered it, he was the first to manufacture it on a commercial scale. At Bathgate "he produced 13 lbs of paraffin wax per ton of cannel-coal plus 30 gallons of a lubricant which proved to be an excellent cart grease, of which he had sold as much as 8000 gallons since he started production. May this declaration help to establish his priority as a manufacturer before the courts". Young's paraffin wax was sold in two qualities with a melting point of 55° C and 60° C, each, so it was realized being a mixture of solid hydrocarbons.

In France and other European countries experiments were carried out to manufacture paraffin wax from other base materials. Much money was lost in Saxony during the years 1856-1858 [3] on trials with unprofitable base materials. In Germany the dry distillation of lignites was only partly successful, it was undertaken on a commercial scale by Wiesmann and Comp. at Beuel (near Bonn) and by Gähler and Comp. in Saxony.

However, paraffin wax could also be made from crude oil. Von Kobell had prepared it from the crude of Tegernsee (Bavaria) and in England "Rangoon tar" was being profitably used to manufacture "Belmontine" named after the Belmont (London) factory. Warren de la Rue and Müller [4] had examined a

---

[1] P. Wagemann, Destillationsprodukte verschiedener Rohmaterialien zur Gewinnung von Photogen und Paraffin (Polyt. J. vol. 145, 1857, pags. 309-310).

[2] C. von Reichenbach, Notiz zur Geschichte des Paraffins (J. prakt. Chemie, vol. LXIII, 1854, pags. 63-64).

[3] O. Buchner, Die Mineralöle (F. Voigt, Weimar, 1864, pags. 9-12).

[4] Warren de la Rue and Hugo Müller, Chemical Examination of Burmese Naphtha or Rangoon Tar (Philos. Magaz. vol. XIII, 1857, pags. 512-517).

A. Albrecht, Das Paraffin und die Mineralöle (Stuttgart, 1874).

Burmese crude and stated that they recovered 96% as distillates when distilling with superheated steam. If the temperature of the steam remained below 160° C no paraffin wax could be separated but in increasing its temperature for the heavy fractions range about 20% of distillate was recovered from which "one-third of solids separate" and a further 21% of distillate was entirely semi-solid. All distillates obtained beyond a temperature of 145° C exposed to a freezing mixture gave "10 to 11% solid constituents on Rangoon Tar".

On December 23rd, 1854 Warren de la Rue had patented a commercial process for the manufacture of paraffin wax: "Improvements in treating products arising from the distillation of a certain Tar or Naphtha to render the same suitable for dissolving or removing Fatty and Resinous Substances" in which a nitric acid was used to remove the aromatics and to convert them into nitro-benzole. Warren de la Rue tried to establish the formula of the solid parts which he believed to be C_nH_{n+1}. His process was soon used by Messrs Price's Candle Company. The raw material was Burmese crude imported in "hermetically-closed, metal tanks to prevent the loss of any constituent".

In Barlow's paper [1] we read:

The processes adopted

"In the commercial processes, as carried out by Mr George Wilson, at the Sherwood and Belmont Works, the crude naphtha is first distilled with steam at a temperature of 212° F; about one-fourth is separated by this operation. The distillate consists of a mixture of many volatile hydrocarbons; and it is extremely difficult to separate them from each other on account of their vapours being mutually very diffusable, however different may be their boiling points. In practice, recourse is had to a second or third distillation, the products of which are classified according to their boiling points or their specific gravities, which range from .627 to .860, the lightest coming over first. It is worthy of notice, that though all these volatile liquids were distilled from the original material with steam of the temperature of boiling water, their boiling points range from 80° F to upwards of 400° F.

These liquids are all colourless, and do not solidify at any temperature, however low, to which they have been exposed. They are useful for many purposes. All are solvents of caoutchouc. The vapour of the more volatile, Dr Snow has found to be highly anaesthetic. Those of the lower specific gravity, called in commerce Sherwoodole, have great detergent power, readily removing oily stains from silk, without impairing even delicate colours.

---

[1] Rev. J. Barlow, On Mineral candles and other products manufactured at Belmont and Sherwood (Proc. Royal Instit. Great Britain vol. II, 1854/58, pags. 506-508).

The distillates of higher specific gravity are proposed to be used as lamp-fuel; they burn with a brilliant white flame; and, as they cannot be ignited without a wick, even when heated to the temperature of boiling water, they are safe for domestic use.

A small percentage of hydrocarbons, of the benzole series, comes over with the distillates in this first operation. Messrs De la Rue and Müller have shown that it may be advantageously eliminated by nitric acid. The resulting substances, nitro-benzole, & c., are commercially valuable in perfumery, &c.

After steam of 212° has been used in the distillation just described, there is left a residue, amounting to about three-fourths of the original material. It is fused, and purified from extraneous ingredients (which Warren De la Rue and H. Müller have found to consist partly of the colophene series) by sulphuric acid. The foreign substances are thus thrown down as a black precipitate, from which the supernatant liquor is decanted. The black precipitate, when freed from acid by copious washing, has all the characteristic properties of native asphaltum. The fluid is then transferred to a still, and, by means of a current of steam made to pass through heated iron tubes, is distilled at any required temperature. The distillates obtained by this process are classed according to their distilling-points, ranging from 300° to 600° F The distillations obtained, at 430° F and upwards, contain a solid substance, resembling in colour and in many physical and chemical properties, the paraffine of Reichenbach; like it it is electric, and its chemical affinity is very feeble: but there are reasons for believing that a difference exists in the atomic constitution of the two substances. The commercial name of Belmontine is proposed for the solid derived from the Burmese naphtha. Candles manufactured from this material possess great illuminating power. It is stated that a Belmontine candle, weighing $\frac{1}{8}$th lb., will give as much light as a candle weighing $\frac{1}{6}$th lb., made of spermaceti or of stearic acid. Its property of fusing at a very low temperature into a transparent liquid, and not decomposing below 600° F recommends this substance as the material of a bath for chemical purposes. As to the fluids obtained in the second distillation, already described, they all possess great lubricating properties; and, unlike the common fixed oils, not being decomposable into an acid, they do not corrode the metals, especially the alloys of copper, which are used as bearings of machinery. This aversion to chemical combination, which characterizes all these substances, affords, not only a security against the brasswork of lamps being injured by the hydrocarbon burnt in them, but also renders these hydrocarbons the best detergents of common oil lamps. It is an interesting physical fact, that some of the non-volatile liquid hydrocarbons possess the fluorescent property which Stokes has found to reside in certain vegetable infusions."

PLATE VIII

Pouring molten paraffin wax into a mould in a candle factory at Puerto Cabello (Venezuela) (Shell Photograph, 1954).

PLATE IX

The traditional Chinese method of candle-making. The candle wicks are suspended from a type of wheel and the string is basted with the liquid wax (Shell Photograph, 1951).

PLATE X

A candle factory about 1849.

Candle manufacture, sorting and cutting to size.

Barlow tells us that "Burmese naphtha is almost destitute of hydrocarbons belonging to the olefiant-gas series, but contains hydrurets or radicals of the ethyle series (paraffins!) and substances of the benzole series (aromatics!)".

The refining process differed little from that used by Young at Bathgate which plant Lunge visited and described [1]:

Young's process for separating paraffine from paraffine oil

"Mr James Young extracts paraffine from the oil prepared as above by cooling it to 30° or 40° F The lower the temperature, the larger the amount which crystallises out. It may be obtained sufficiently pure for lubricating purposes by merely filtering off and squeezing out fluid impurities from the mass by powerful pressure.

The paraffine may be purified further by alternate treatments at about 150° F with oil of vitriol and caustic soda. The treatments with acid are to be continued until the latter produces no more blackening. The solid hydrocarbon is then to be washed with caustic soda until all acid is removed, and then with boiling water. The treatment with boiling water should be performed several times.

The oil from which the paraffine has been removed by exposure to cold is by no means freed from the whole of the solid; it is, in fact, a saturated solution of paraffine at the temperature to which it was exposed. It is sometimes advantageous, before extraction by cold, to concentrate the paraffine in the paraffine oil, by subjecting the latter to distillation, until one half or two-thirds of the fluid has distilled over; by this means the yield of paraffine is proportionately increased.

The amount of solid matter distilling over with naphthas may be seen by consulting the results obtained by M.M. Warren de la Rue and Hugo Müller, in their fractional distillation of Rangoon tar. It is to be observed that solid hydrocarbons differ in the degree to which they pass over with the vapour of fluid hydrocarbons. Thus while pyrène and chrysène only appear among the very last products of the distillation of coal at high temperature, naphthaline will often distil over at very moderate temperatures in presence of volatile fluid hydrocarbons. The author of this article has repeatedly seen considerable quantities distil over in a current of steam at the pressure of the atmosphere, and consequently at 212°. The facility with which solid hydrocarbons pass over in the vapour of volatile fluids, depends not only upon their boiling points, but also to some extent upon special tendencies varying with the nature and state of admixture or combination of the substances operated on."

[1] Dr. Lunge, Fabrique de paraffine de Young à Bathgate (Ann. du Génie Civil vol. 7, 1868, pags. 57/59).

He also tells us that Young recovered about 1% of paraffin wax on Boghead Coal which he sold in cakes to candle manufacturers such as Price. Soon the paraffin wax produced by the Scottish shale industry was no longer able to cope with the demands of the candle manufacturers, and waxy crudes began to dominate the market. From 1857 onwards Messrs Price regularly received shipments of Rangoon Tar, first in wooden barrels, then in metal containers to avoid evaporation of the valuable low-boiling fractions.

Warren de la Rue himself [1] has described the process in detail. The Burmese crude was distilled in cylindrical stills provided with perforated steam coils, the steam being introduced when the temperature of the liquid had risen to 100° C. 16-20% of the crude was first taken off to be redistilled and refined. The residue was then again subjected to steam distillation with steam temperatures of 315-370° C. The waxy distillate was redistilled and cooled, the solids separated and pressed cold and hot and treated with an equal amount of sulphuric acid at 100° C. The wax was then ready to be used for candle manufacture or for the lubrication of machinery. In some cases the waxy distillate was given an acid wash before redistillation.

A similar flow-sheet was followed on the Continent where waxy Galician crudes were beginning to be used for the manufacture of paraffin wax [2]:

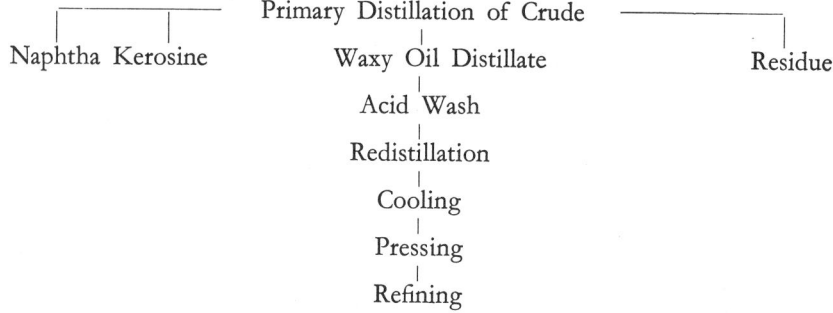

The difficult separation of oily parts and solid paraffin wax from slack wax or crude wax was not solved properly until Hodges introduced the "sweating process" in 1871. A simple treatment in hydraulic presses proved insufficient and it became common practice to give the crude wax from the cool-house an acid wash and then to add about 25% gasoline to the molten wax, which mixture was then left to solidify in basins before it was pressed in two to

---

[1] Warren de La Rue, Distillation des naphthes et des goudrons (Le Technologiste, vol. XX, 1859, pags. 352-354).

[2] E. Davies, J. M. Syers and C. Humphrey, Refining paraffin wax (Le Technologiste vol. XX, 1859, pags. 254-255).

three steps or more if need be. The last traces of naphtha could easily be removed by heating at 120° C and the gasoline needed could be obtained from the same Burmese crude as the paraffin wax.

It was soon realized that the paraffin wax produced from shale-oil or coal-oil needed special treatment [1]. Pre-refining the crude wax demanded at least 10% sulphuric acid. The pressed wax was often treated with a 50% sulphuric acid and washed before the second pressing and then refined after the last pressing by first mixing it with ½% of stearin and treating it with a 70% acid. After a water wash it was then again mixed with ½% of stearin and boiled with 1% of a 40° Bé caustic soda, after which treatment the dirt settled from the molten wax which was then cast into slabs.

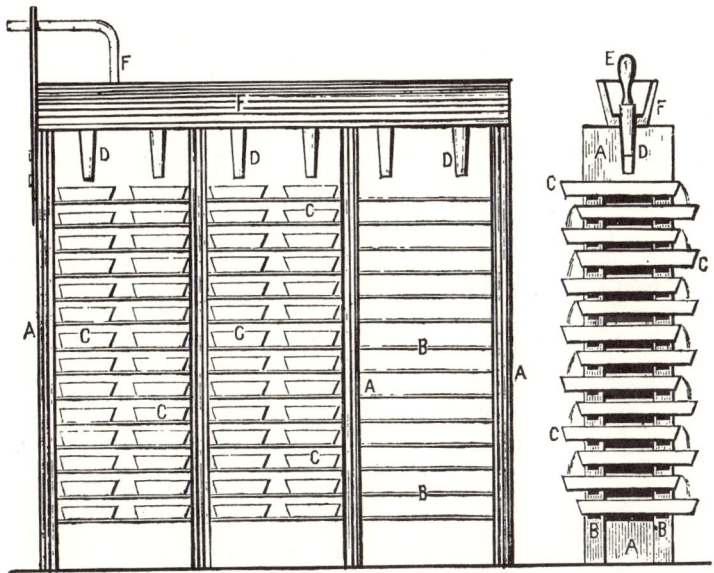

Fig. 17. Binet's rack for casting refined wax into slabs for shipping.

Such a long series of operations seemed necessary to remove the traces of asphaltic compounds from the waxy distillate and the crude wax which not only impeded proper distillation but prevented the production of a perfectly white wax. Not until better fractionation and dephlegmation of the distillates

[1] C. G. Müller, Über die Reinigungsweisen des Rohparaffins (Dinglers Polyt. J. vol. 154, 1859, pags. 227-232).

was achieved could the manufacture of wax be simplified. It remained a time-consuming, expensive process consuming a considerable amount of chemicals and hence involving heavy losses of material.

The very clumsy way of refining the pressed wax was, however, soon simplified. The decolourizing power of vegetable carbon was discovered by Lowitz in 1790, but Figuier, an apothecary of Lyons, soon found that animal carbon such as bone-black was even superior. This product was improved by the work of Payen, Bussy and Desfosses [1].

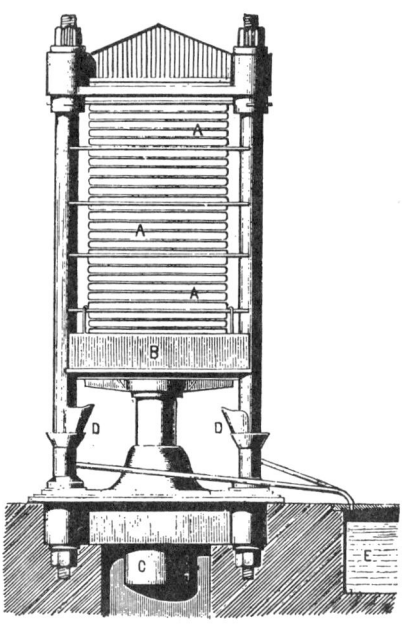

Fig. 18. Hydraulic press used for the refining of wax with gasoline.

If properly dried at 110 to 150° C this bone-black was eminently suitable for the decoloration of pressed wax. This process was introduced by H. Perutz at Pest in 1868 [2]. About the same time Hochstetter prepared another

---

[1] Ann. Chem. Pharm. vol. 55, page 241; vol. 59, page 354; Dinglers Polyt. J. vol. 9, page 206; vol. 27, page 372.

[2] H. Perutz, Industrie der Mineralöle (Gerold, Wien, 1880, vol. II, pags. 111-114).

J. Fordred, F. Lambe & C. Sterry, Verfahren zur Reinigung des Petroleums, des Paraffins und anderer Kohlenwasserstoffe (Dingler's Polyt. J. vol. 193, 1869, page 437; Armengaud's Génie Industriel, March 1869, page 156).

L. Schuch, Zur Reinigung von Paraffin (Dtsche Industrie Ztg. 1869, page 208).

decolourizing powder as a by-product of the manufacture of Prussian Blue and Turnbull's Blue. The complicated refining of pressed wax was now reduced to an acid wash with 5 to 10% of sulphuric acid, sometimes still followed by a stearin-caustic soda wash to collect the dirt in soap-form and a water wash. Usually, however, this operation was entirely obviated and the molten wax was refined with 2 to 3% of bone-black and filtered. In the case of very yellow waxes the amount of bone-black went up to some 10% but the entire operation was now reduced to $\frac{1}{2}$ to 3 hours.

Very soon paraffin wax became the main ingredient of candles and tables were published for mixtures of shale wax and oil wax with stearin in order to produce the proper mass with a melting point of 45 to 46° C [1]. Even today a very large proportion of paraffin wax is still used for the manufacture of candles and every country has its own specifications for the colour and appearance of such candles.

Two new types of waxes date back to the early phase of the oil industry. In 1874 Chesebrough took out a British patent for the manufacture of "vaseline". A waxy residue of 45 to 47° Bé was heated for 24 hours in open vessels and filtered over carbon black, becoming pure white and salve-like. It was recommended for unguents and cosmetics [2].

More attention had also been given to the natural waxes and Ujhely had discovered and patented on June 13, 1871 a process to decolorize ozokerite by bleaching and treatment with bone-black only. Together with a number of bankers he founded the first Austrian Ceresine factory at Stockerau. Similar patents were taken out by Pilz of Carlsbad and Otto of Frankfurt who later also succeeded in producing commercial products. Pielsticker treated ozokerite in the old-fashioned way with hot sulphuric acid, followed by a water wash, a treatment with barium carbonate and caustic soda and a filtration through clay [3]. Oxygen compounds were found in these natural waxes and it was suggested by several inventors that such substances as pyropissite might serve as a base material for the manufacture of fatty acids, but nothing came of such processes.

Further applications of paraffin were announced. Monet [4] propagated its use as "a lubricating substance unalterable at 300 to 400° F in engines such as hot-air machines, as it is also an infallible preservative against oxidation".

[1] M. Grotowsky, Point de fusion des mélanges de paraffine et de stéarine (Le Technologiste vol. XXXI, 1870, pags. 371/372).
[2] R. A. Chesebrough, New product from petroleum (Brit. Patent no. 1012, March 23, 1874).
[3] C. Pielsticker, Refining ozokerite (Brit. Patent no. 797 of Febr. 25, 1876).
[4] M. A. Monet, Paraffin for Oiling at a High Temperature (J. Franklin Instit. vol. 55, 1868, page 83).

On April 3rd, 1865 a paper [1] was read to the Pharmaceutical Society, Edinburgh on a process invented by Dr Redwood. Meat immersed in a bath of paraffin at about 250° F rapidly lost air and water leaving the juice of the meat in a concentrated state. Redwood's Patents Company were stated to be experi-

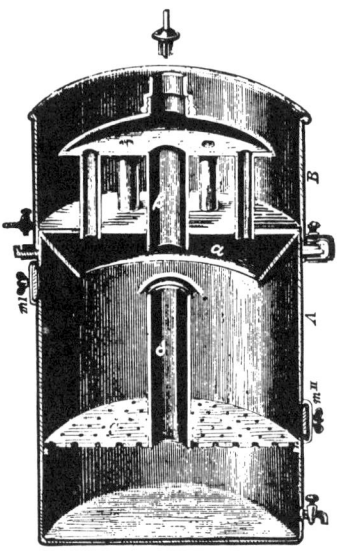

Fig. 19. Thorn's apparatus for the extraction of crude ozokerite.

menting from 1864 onwards on bacon, butter, eggs, sausages, cheese, hams, etc. It was shown that it was perfectly easy to remove the paraffin before cooking. Both applications were actually adopted.

Actually by 1860 the manufacture of paraffin wax had already made a most promising start and most of the basic processes still used in the wax factory had been discovered. The paraffin wax candle was on its way to success and the physical and chemical principles underlying its use were already most fittingly described by Michael Faraday in his successful Christmas Lectures to the Royal Institution, a booklet which deserves reading even in our age of electric light [2].

---

[1] J. Mackay, A New Method of preserving Beef, Mutton and other Animal Substances (Pharmaceutical J. vol. VII, 1865/66, pags. 560-561).

[2] Michael Faraday, The Chemical History of a Candle (edited by Sir W. Crookes, London, 1861).

CHAPTER EIGHT

NEW BYPRODUCTS FOR BURNERS, BEARINGS AND BITUMENS

No more striking contrast between the young American petroleum industry and its European contemporary can be obtained than by turning to their use of the residue left over after manufacturing the naphtha and kerosine. As the American refiners very often "distilled to coke" they had little residue to bother about. In this "one-crude country", Pennsylvania, much less attention was paid to the manufacture of a range of lubricants, asphalts or fuels, which in the case of European refineries often helped to combat the strong competition of other lamp-oil producing industries such as the shale-oil, the lignite, the boghead coal, and the coal-tar industries. In Europe the production of the maximum amount of marketable products was essential. In Pennsylvania the rapidly swelling stream of oil wiped out its competitors by mere volume and a complete efficiency of its handling was not so essential.

Thus Thomas Gale [1] can still grow lyrical on its medicinal values and recommend this application of both crude oil and residue in these glowing words:

"Hundreds in all parts of the land are asking: "What is it good for?" The answer in general is, that it is good for the same that other oils are good for.—Some of its specific uses and excellencies will be pointed out.

1st. It possesses medicinal virtues. Seneca Oil has been known, for a long period, to operate very efficaciously as a topical application in burns, bruises, sprains, sores, fresh wounds, etc. It penetrates, purifies, soothes and heals. Every farmer should keep it to use for galls, scratches, lameness and like injuries among his horses and cattle. For such purposes it has never been surpassed. It will cure if anything can. And should it fail to remove at once, soreness and pain, as all remedies sometimes will, it will alleviate. Some who have tried its soothing, restoring power on their burnt fingers, frosted feet, chilblains, etc., insist that no family should be without it. As Seneca Oil, it has long been sold at a dear rate. This indicates a widespread assurance of its efficacy in the line of working off pain and forwarding a healing process. It is further beneficial taken in doses, and also applied outwardly in colds, and where the lungs and bronchial membrane are somewhat affected. For not more readily does

---

[1] Thomas A. Gale, Rock Oil in Pennsylvania and Elsewhere (Erie, 1860, pp. 49-51).

light permeate glass, than this penetrating liquid finds ingress to the irritated chest, there to diffuse its balm. Let the hoarse, whispering, wheezing sufferer swallow it freely, rub it on the surface and dry it in thoroughly, and soon he will be able to draw a long breath again, and to talk with all the ease and garrulity of age. Cases might be cited where those of consumptive tendencies are taking it instead of Cod Liver Oil, with happy results.

It has long been used to allay violent pain in the stomach and to check diarrhea. But the question of its power to control disease in such circumstances, as well as in others, must be refered to the Dons of the healing art.

The United States Dispensatory says: "Externally Petroleum is employed as a stimulating embrocation in chilblains, chronic rheumatism, affections of the joints and paralysis. It is an ingredient in the popular remedy called British Oil".

"Petroleum is accounted a stimulating antispasmodic and sudorific. It is occasionally given in disorders of the chest, when not attended with inflammation. In Germany it has been extolled as a remedy for tape-worm —one part of Petroleum, and one and a half parts of tincture of assafedita".

A medical friend and skillful practitioner gives it credit for possessing valuable alterative and cathartic properties; for being a good expectorant; also beneficial in certain cutaneous diseases. He further speaks of its soothing influence in severe colds on the mucus membrane of the air passages. This much is so; when given to a croaking boy who, as some would word it, was completely sewed up, it worked like a charm. Try it on barking children. But we leave it to the profession to discuss its remedial virtues, and so far as they see fit give it a place in their saddle-bags."

Crude oil or oil fractions were used on a modest scale for the enrichment of town gas, on which carburetting operation Küchler [1] wrote an excellent handbook. Many inventors too were trying to construct a motor running on a petroleum fuel [2], but none of them had made this application a lasting success yet in the period 1860-1880 which we are discussing. Generally speaking the cost of the fuel was still prohibitive though Abland proudly compares his motor with the "Lenoir Gas Engine, the Siemens' Motor and Lehmann's Hot-Air Engine" and claims that his $1^3/_4$ HP motor consumes about 6 cents worth of petroleum per hour!

However, considerable developments took place by the invention of burners suitable for crude oil and residues [3], which turned a waste product into a major fuel for traffic on sea and rail.

[1] F. N. Küchler, Handbuch der Mineralöl-Gasbeleuchtung (Berlin, 1878).
[2] J. W. Young, Notes on the Practical Development of the Oil Engine (Trans. Newcomen Society, vol. XVIII, 1936/37, pp. 109-129).
Yves le Galliec, Les origines du moteur à combustion interne (Techniques et Civilisations, vol. II, 1951, pp. 28-33).
J. Hock, Petroleum Motor (J. Soc. Arts, 1874, p. 798).
W. E. Abland, A Petroleum Motor (J. Franklin Instit., vol. XCVIII, 1874, pp. 87-89).
[3] H. A. Romp, Oil Burning (The Hague, 1937).

1. *Burning a New Fuel Oil*

Attempts to construct burners suitable for this new fuel were made very much independently in the U.S.A., in Western Europe and in Russia, because of the rather primitive communications of those days. Only after 1880 did the efforts of these regions become more co-ordinated.

The most rapid development took place in Russia where large amounts of astatki ("waste oil") were available as soon as refining started in Baku, for the heavy crude had but a relatively small yield of kerosine. This seemingly valueless waste oil was turned into valuable mazout (fuel oil) when the Russian inventors constructed burners on the principle of pulverizing the residue with steam by means of devices, which were mostly of the slot type. This led to rapid practical success, for the mixing of steam and oil not only preheated and pulverized the oil, but the steam made it possible to burn very heavy fuels without soot formation by chemical action. This "water-gas reaction" must have been common knowledge in Russia before 1860. By that time the so-called Astrakan burner was used for heating houses. This burner consisted of one or more flat dishes placed in the bottom of a stove, into which oil flowed down from a reservoir by gravitation. The oil was heated and evaporated by radiation from refractory bricks placed over the oil flame. As the combustion of the heavy oil led to heavy soot formation, water was introduced into the dishes from a second reservoir, where it evaporated spontaneously and threw up the oil in drops thus contributing to a much less smoky combustion.

One of the first inventors using this principle for a burner was Spakowski who in 1866 patented an oil burner which he improved in 1870. The oil was admitted to the centre of this burner and surrounded by a concentric ring of steam, which apart from propelling the oil also drew in the air. By mere chance it gave a flame very suitable for marine boilers avoiding damage by impingement of the flame on its surface. This was the first "nozzle type" oil burner.

In 1870 Lenz designed what was the first "slot type" oil burner, which came to be known as "drooling burner". It had an upper slot for the oil, a lower one for the steam, which were arranged in such a manner, that four flat streams of oil dropped on to a flat stream of steam.

By 1876 a third type, the "tube type" oil burner was developed by Körting. This was a forerunner of the Venturi burners, frequently using steam injection for the supply of air necessary for the combustion. This combination of oil pulverisation and air suction promoted a thorough preliminary mixing of air and steam, and hence facilitated contact of air and oil.

Hence by 1880 the Russians disposed of several types of good oil burners [1]:

1. The nozzle type (Spakowski, Urquhart)
2. The slot type (Lenz, Brandt, Artemev)
3. The tube type (Körting)

which began to make fuel oil an excellent competitor of coal all over Russia. By 1870 Russian engineers were already dreaming of using fuel oil in the Ukraine factories and on the Black Sea, but they complained that transporting oil in barrels made it too expensive in Odessa [2]. In 1870 an engineer, Porjetzki had proposed using it for locomotives, an idea which was applauded in Government circles. The "Russian Society for Steam Navigation, Commerce and Railways" conducted trials with great success. The steamer which was to sail from Odessa to Nicolajeff was held up by heavy storms but on land fuel oil certainly proved equal to coal, the saving in fuel amounting in most cases to 50%. It was then proposed to try and fire the boilers of the Volga steamers with fuel oil, for this was to lead to a saving of 300 Rubles per ship per day.

By 1874 this had been realised and almost every ship on the Caspian Sea and the Volga including naval vessels was running on oil. Their burners were all of the steam-pulverizing type. The steam was obtained from low pressure boilers (5 atm.) fed with salt water, the salt content of the boilers being kept down by temporarily blowing off. By about 1890 nearly all Russian locomotives, except those in the coal-producing districts and in Siberia, were using fuel oil. The Russian navy was far ahead of that of other countries, where fuel oil was first introduced in 1893 (Italy), 1895 (Great Britain) and 1910 (United States of America). This late general use of fuel oil was intimately linked up with an unsuccesful start with unsuitable burners. This is very evident in the case of the United States. The American inventors started constructing burners on the theoretically correct principle that the oil should be evaporated before the oil gases were to burn. This is what actually happens in our modern burners, but the American inventors struck out on a different path from the Russians. They used a separate space to heat and evaporate the fuel oil, the gases of

---

[1] Bloomer, Petroleum Fuel (St. Petersburg, 1888).
Gulischambarow, Die Naphthaheizung der Dämpfer und Lokomotiven (St. Petersburg, 1894).
Lew, Die Feuerungen mit flüssigen Brennstoffen (Leipzig, 1925).
Nobel, On the extended Use of Oil Residues in Furnaces without Spraying (Moscow, 1882).
Stepanoff, Die Grundlagen der Lampen Theorie (Leipzig, 1906).
Teleczinski, Petroleum and its Use as a Domestic and Commercial Fuel (Lemberg, 1870).
Versenev, The Management of Petroleum-fired Boilers (Russian, Moscow, 1891).

[2] Anon., Petroleum in Russland zur Heizung auf Schiffen und Lokomotiven (Dtsche Illustr. Gewerbe Ztg, . 1871,No. 33, p. 259).

which were to be led under the boilers. Thus John Edward Biddle's burner of 1860 consisted of a cast-iron bucket or basket of glowing anthracite to evaporate the oil dripping in it from a pipe. Neither this burner nor others consisting of series of channels through which the oil flowed (1862) were a success.

Thomas Gale had great expectations of the future of fuel oil:

"The astonishing consumption of fuel by the numerous Steam Engines on our lakes, rivers and Railroads, suggests another function of rock oil—that of generating steam. The number of these in motion, vast as it now is, within no very extended period, will be doubled and quadrupled.

Where will fuel be found to produce the oceans, in the aggregate, of steam needed daily? Once wood sufficed, to heat the few boilers steaming on our Railroads. But years ago in many places the supply failed. Recourse was had to coal, first anthracite and then bituminous. Some predict that soon coke on locomotives in this country will take the place of coals, as it already has come into this use in England, All these coals, generating steam by heat instead of blaze, are bulky and heavy. Rock oil possesses in these respects manifest advantages. One gallon of it will raise more steam than many times its weight and volume of coal or coke—no small consideration where the locomotive and steam-engine must carry their fuel a great distance. It admits of being more readily placed under the boiler, than the fuel in use can be. It can be furnished for that purpose in our great cities, by the quantity, at a low figure. Whether it will best be applied by itself, or mixed with other materials, is a discovery to be evolved in the unknown future.

A very distinguished machinist writes me on this subject as follows: "Rock Oil crude, can be used as a portable fuel, and also as a fuel for generating steam on boats and locomotives; and would be a very economical fuel for that at a cost not exceeding 30 cents per gallon; and as I think will come into general use for that purpose".

Since the above was written the extract below appeared in the Titusville Gazette. "A citizen of Pittsburgh of great scientific attainments and great inventive genius, has hit upon an idea which may, if carried out, cause an active demand for all the oil that can be produced. His plan is to construct steamers with the engines so arranged that the boilers may be heated from the burning of oil instead of coal. The calculation is that the weight of the oil would be six-sevenths less than the coal now used. This would vastly increase the carrying capacity of our ocean steamers, as well as create an almost illimitable demand for that which is so much exciting people in the neighborhood of Titusville".

According to this, one-seventh of the hundred pounds or tons required to drive an engine with coal, would be adequate, if rock oil were substituted.—What is wanting is to have the thing tested. We known the most scientific men are subject to reverie; their fancies they pronounce feasible and are ready to worship them; while at the same time with philosophic skill they expose the visionary incongruities originated by others. We have seen cultivated minds cautious and regular usually in their movements, dart off suddenly in a course as eccentric as the orbit of a comet.

Not that the proposal above mentioned is ideal; good judges say otherwise; and all we mean is that it has not been tried. It looks practicable. And if the ingenious gentleman in Pittsburgh, or any other can get up a first rate apparatus for generating steam from this oil, with the saving of fuel and tonnage claimed, his patent will be worth more than the richest well in Oildom.

A friend, since the line above was penned, informs us he has entered in the Patent Office a caveat against such a patent being granted, thereby securing to himself the rewards of his own fertile genius. His plan is by means of pumice-stone, perforated plates, etc., to feed the oil slowly and distribute the heat evenly under the boiler, taking care to preclude waste from draught.

It has been said 1 ton of coal oil will evaporate as much water (the fire-box, boiler, etc., being properly constructed) as 4 tons of coal. Suppose rock oil to do as much. The steamer which now consumes 50 tons of coal daily, could steam four days with that weight of fuel, if it were rock oil. What a saving of space for passengers and cargo! The economy of oil for fuel appears to best advantage in the case of steamships crossing the ocean and obliged to lay in before starting 1,000 tons for the voyage. By substituting rock oil, only 250 tons of fuel are required, leaving room for 750 tons of goods to be transported. Surely such economising of space is no trifling matter.

When the much smaller number of tons of oil which a vessel would be required to ship for fuel, was suggested to Commodore S., of Washington, he remarked, that providing the cost were twice as much, oil would be adopted, on account of its great advantages. If adopted, he remarked the tall chimneys, 7 feet in diameter, are unnecessary. These are always in the way of the sails, and in a storm or blow, render the danger greater. In these views the chief engineer concurred.

We may very soon see steam-ships starting for a distant port, with oil as the steam-generator. By touching at Trinidad and dipping into the lake of oil there a relay of propelling forces will always be at command. Oil wells may be pumped dry; that deposit scorns the idea of exhaustion."

The inventions of Salisbury, Roger, Sadler and Orvis were soon forgotten and Walker's burner of the "scent-spray type" had but some temporary success. In 1865 three engineers of the U. S. Navy, Wood, Whipple and Stimmers had carried out trial runs with Linton-Shaw burners which showed how oil fuel held great promise in easy handling and quick heating up of the boilers [1]. They calculated that the small amount of space required for the oil would mean a saving of $ 15,000 on a return-trip to Europe. Unfortunately some of their contemporaries were of a very different [2] opinion:

"The absolute heating power of petroleum is 1,835 times greater than that of a pound of anthracite. The heat produced by petroleum costs over seven times more than the heat produced by coal. It is probable that a steam vessel which requires, say 24 men in the fire-room, could dispense with at least 14 of them by the use of petroleum. Experiments carried on with the machinery of the U.S. gun-boat Palos in Boston harbour show that the result is still further against the economical use of the oil as a steam fuel. Cost of the apparatus is very high and it is evident that

---

[1] Anon., Emploi du Pétrole comme Combustible pour les chaudières Marines (Revue Univ. des Mines, vol. XVIII, 1865, pp. 220-221).

[2] Anon., Petroleum Fuel for Steam Ships (Chemical News, vol XVII, 1868, pp. 224-225).

portions of it will require constant renewal. The pipes and passages became so choked with soot and carbon in less than 48 hours that fire went out, and could only be renewed by taking the fixture apart and cleaning it".

The gravest danger, however, was that of storing such inflammable oil on board:

"These facts point out at once the extraordinary care that must be taken in storing this substance on board ship, in order to guard against accidents of the most frightful character. It therefore seems clear that a proper regard for safety demands that the tanks containing the petroleum should be immersed in water; when the weight of these tanks and cisterns is borne in mind, together with the bulk of petroleum and coal composition (a cubic foot of the former weighs 50 lbs., and a cubic foot of anthracite, as stored in bunkers, weighs 52.5 lbs.), and also their relative heating values, which may be set down as 1.4 for the former and 1 for the latter, it may be with safety assumed that there would be a saving in weight and space of not over 30 per cent. in stowage capacity by substituting the oil for the coal. It is therefore evident, from what we have said, that even assuming that the oil and the coal can be burned with equal facility, and with an equal degree of liability to derangement, in steam-boiler furnaces, the excess of the heating power of the petroleum over that of the oil is so very much less than its excess of cost, that there is not the slightest probability, as long as these ratios exist, of petroleum ever taking the place of coal as a steam fuel.

When we look at the great complication and danger that must be added to steam machinery in order to burn the oil, and the liability to derangement and the want of durability, it is not likely that any prudent steam navigation company would allow it to be employed in their vessels, even if they could find an engineer to recommend it."

This unsatisfactory start of burner design and unsuccessful experiments gave oil burning a bad reputation in the United States and therefore we have to wait until the last decade of the nineteenth century to find merchantmen on the Pacific Coast giving the new fuel a trial. Nearer Pennsylvania the Pennsylvanian Railway Company provided some of their engines with oil burners of the Russian type in 1887, but it took many more years before this fuel was generally used on locomotives even in the oil-producing regions. The man who put the inventors on the right track was Commander Isherwood of the U. S. Navy who in 1875 studied the possibilities of the new fuel for the Navy and concluded: "Atomisation is the only way" of burning oil properly, though he never suggested in his report how this was to be achieved in the burner!

In France [1] scientists and engineers co-operated at a much earlier date to

---

[1] Colomer et Dordier, Les Combustibles Industriels (Paris, 1921).
Roszmäler, Die flüssigen Heizmaterialien (Wien, 1910).
Siebert, Petroleum-heizung und Rohöl-feuerungen (Dresden, 1921).
Syndicat d'Applications Industrielles, Les Combustibles Liquides et leurs Applications (Paris, 1921).

make fuel oil a success. This came about because of men like Sainte-Claire Deville, Audouin and Dupuy de Lôme, who all had been working on the gas-lighting of Paris, and who were directed to the study of fuel oil at the Paris Exhibition of 1867. Here several inventors demonstrated applications of petroleum products. One, Levêque, had tried to show his process for the heating of boilers with evaporized petroleum, but the Committee forbade this because of the risk of fire [1]. Demonstrations were allowed in the laboratory directed by Henri Sainte-Claire Deville and there Audouin demonstrated his "burner". Paul Audouin (1835-1912) had made attempts to use the heavy residues of coal-tar and he had constructed in Juli 1865 the prototype of his "grate". In 1867 this had taken the form of a number of vertical fire bars covered with a film of oil evaporated by the radiant heat of the flame itself; a year later he had made another type (together with Sainte-Claire Deville) with inclined fire bars and adjustable air registers [2] to give the highest possible velocity to the air at the point of evaporation of the oil film. This burner gave excellent mixing, clean and efficient combustion and it rightly attracted the attention of Napoleon III during his visit of the Exhibition. The emperor ordered Deville to pursue studies of a mineral fuel which would serve as a substitute for the coal used on board his yacht Puebla, the soot and smoke of which annoyed his guests.

Audouin's description of his experiments shows that he had tried to solve the problem of getting rid of the complicated and expensive American oil burners of his day. Hence in his burners he introduced natural draught as the force moving the flame combined with film evaporation of the oil. He paid attention to the air velocity as a means of evaporating the oil and he reduced the resistance to the air to the utmost. Above all he was aware of the importance of minimum excess air in order to achieve the highest flame temperature, perfection of combustion and efficiency. Like Knab [3] he had a good theoretical knowledge of the phenomenon of combustion and he preferred liquid fuels to coal-tar or petroleum as they required less handling, were easy to store and left hardly any ashes.

Deville himself, though more interested in scientific research such as the calorific value of different fuels, had run an 8 HP engine on oil at the boiler

[1] Rapports du Jury Int. Exposition Univ. de 1867, Paris (Vol. V, Paris, 1868, pp. 68-93).
[2] P. Audoin, Application des hydrocarbures liquides à l'obtention des hautes températures et au chauffage des machines à vapeur (Ann. Chimie Phys., vol. (4) XV, 1868, pp. 30-40).
Brevet No. 77.383 dated 3.8.1867 later ammended on May 7 & June 9, 1868 and 15.11.1871.
[3] Cl. Knab, Sur l'emploi des combustibles de l'huile minérale pour le chauffage des navires à vapeur et sur l'application de l'oxygène à la combustion (Ann. Génie Civil. 1868, No. 5, pp. 305-321).

plant of Belleville and found that it compared favourably with coal in such practical trials. With Audouin he now applied the "Deville grates" [1] to different types of engine. Dieudonné helped to conduct such trials on locomotives [2] and Dupuy de Lôme showed that they could be used successfully on board the Puebla with its 60 HP engine. Ships equipped with such grates could, so they reported, run on coal when crossing the ocean and return on cheap oil bought in the States. The condensed water problem was solved, for this could be recovered from the combustion gases, so they claimed. A miniature "Deville grate" could even be used to heat a laboratory muffle furnace.

The 1870 war put a stop to this development in France and only in 1883 were such experiments resumed. By then friendly contacts with Russia had been established and Russian oil burners became known in France. Hence the Jensen burner, now the standard burner of the Russian Sea fleet, was introduced in the French Navy. The problem of the extra water needed for oil burning on board prompted d'Allest to design compressed-air pulverizing burners which, together with the Guyot burner of the same type, helped to convert the French Navy into an oil-burning fleet.

In Great Britain [3] a good start was made notwithstanding an erraneous concept of the function of the steam in a burner. These British inventors, who mostly used superheated steam, believed that the steam dissociated by the heat of the flame was able to produce extra heat by the combustion of the dissociation products. This use of superheated steam had success [4] because it heated the oil better, reduced the steam consumption with equal velocity effect due to its larger specific volume and because it gave a more steady flame as condensation of water drops was avoided, which water at times even extinguished the flames of these older burners.

Aydon, Wise and Field carried out their first experiments (1865) with superheated steam in South Lambeth, using a burner in which the oil was dripped perpendicularly on to a jet of steam. Aydon himself reported on these experiments [5] in these words:

---

[1] H. Sainte-Claire Deville, C. R. LXVIII, pp. 352 ff.
[2] H. Sainte-Claire Deville, C. R. LXVIII, pp. 353 ff.; LXIX, pp. 933 ff.
[3] Bell, Petroleum Oil Fuel (London, 1902).
Booth, Liquid Fuel and its Combustion (London, 1902).
Butler, Oil Fuel (London, 1914).
Lewes, Oil Fuel (London, 1913).
North, Oil Fuel (London, 1905).
Richardson, Petroleum as a Fuel (London Mechan. Magazine, Dec. 1864).
Thwaite, Liquid Fuel (London, 1887).
Williams, Fuel, its Combustion and Economy (London, 1879).
[4] J. Franklin Instit., vol. 83, 1867, p. 84.
[5] II. Aydon, Liquid Fuels (Engineering, March, 1, 1878, p. 168).

"It was stated that apparatus specifically adapted for the combustion of liquid fuels, which comprised every class of fluid hydrocarbons, might be ranged in five classes. The leading principle of their action was either the subdivision of the liquid as spray, or by percolation through a porous bed, or by preliminary conversion into vapour—when the fuel was mixed with air, or with air and steam, by the instrumentality of jets of steam or of compressed air, or it was burned simply as gas in jets. The earlier system of Mr. C. J. Richardson, in which the liquid fuel, mixed with heated air, percolated upwards through a porous bed, was tried at Woolwich Dockyard, but the performance was not satisfactory, for black smoke and soot were discharged in such abundance as speedily to choke the flue-tubes and stifle the draught. By a subsequent improvement, in which a mixture of steam was introduced with the fuel, a much better performance was effected—the quantity of water evaporated having been increased from $6\frac{1}{2}$ lb. per pound of fuel to from 7 lb. to 18 1/3 lb. per pound of fuel, though the formation of dense smoke was not prevented. The performance of coal under the same boiler amounted to an evaporation of 8 lb. of water per pound of coal. The system of Messrs. Simm and Barff, in which the liquid fuel was vaporised in a retort placed in the furnace, and burned in jets, was tried in 1866, on board the yacht Minnie. The quantity of oil consumed amounted to one-third only of the corresponding quantity of coal. The system was afterwards tried with the addition of steam, and with better results, as the intensity of the combustion was increased, and smoke was prevented.

In the fourth system, patented by the author, in conjunction with Mr. Wise and Mr. Field, in 1865, the liquid fuel was summarily vaporised, by the injection of the liquid into the furnace by the instrumentality of steam, which might be superheated, the supply of air for combustion being at the same time drawn in as an induced current. By this plan, the materials could be instantly and thoroughly mixed and converted into vapour or gas before ignition took place. No alteration of the ordinary furnace or grate was needed, so that either coal or oil could be used. For burning oil, the grate-bars were covered with thin fire slabs and a few cinders; and the ash-pit doors were closed, to keep out surplus air. In March, 1867, this method of burning liquid fuel was tried at the works of Messrs. J. C. and J. Field, South Lambeth, in a Cornish boiler of 20 or 22 horse power, 5 ft. 6 in. in diameter, with a 3-ft. line. The results of several days' experiments showed an average of $19\frac{1}{2}$ lb. of water evaporated per pound of liquid fuel. The boiler previously evaporated $6\frac{1}{2}$ lb. per pound of Aberdare coal. Similarly experiments with a double-flue Galloway boiler, at the chemical works of Mr. Barnes, at Hackney Wick, gave a net evaporative performance of 25.3 lb. of water per pound of fuel. Experiments had been made with other boilers, in which the evaporative efficiency of the liquid fuel ranged from $1\frac{1}{2}$ to 3 times that of coal. Equally good results in favour of liquid fuel were obtained from its employment under a marine boiler at Woolwich Dockyard. The fifth system enumerated, the invention of Mr. Dorsett, in which the liquid fuel was vaporised in a separate boiler or retort, to be burned as a gas, was tried in 1868, at the chemical works of the inventor, at Deptford, and also on board the Retriever steamer. The results in favour of liquid fuel showed a reduced consumption in the ratio of 2.5 or 2.7 to 1 as compared with coal; but against this economy had to be placed the cost of the separate generators and their furnaces, and of a force-pump. The retort, too, was liable to explosion, in consequence of the deposit of solid carbon within it."

Aydon mentioned that he had also used his burner successfully for the reduction and smelting of refractory iron ores in Canada, containing 32% of titanous sand. Although liquid fuel might be burned without the employment of steam, yet "it was consumed most economically, and with the best results, in the presence of steam; and, of course, the more highly superheated the steam, the better was the performance".

The British Admiralty was most interested in these experiments and Admiral Selwyn, who had designed himself a fairly successful burner, got into touch with Aydon and together they developed the Aydon-Selwyn burner, an ancestor of the later Holden steam-jet oil burner (1883). However, oil burners were to be introduced first on the railways, which often had their own coal-mines and gas-works and thus disposed of a steady supply of tar-oils recovered as by-products. Though the Navy obtained good results several factors retarded the introduction of fuel oil. The price of fuel oil in Great Britain was still fairly high and coal-mining interests raised a violent opposition to mineral oil being used as a substitute for coal. They pointed out that fuel oil might not always be available to the Navy, which had a worldwide series of coal-bunkering stations but no equally widely distributed chain of fuel supplies. Moreover the steam consumption of the burners was still a serious drawback in the oil burner which was to become a success only after Körting introduced the so-called pressure-atomizing burner in 1902.

2. *Lubricating the New Machinery*

During the period 1840-1850 the Industrial Revolution was gaining momentum in Europe and the number of machines installed soon begun to increase the power used per individual worker. With this industrialisation came an ever-increasing number of specific lubricating problems. Here lay a field for petroleum products. By 1860 the European refiners were becoming slowly aware of this new market. They still sold a topped residual oil as a general lubricant, though they were already selling some of the heavier-than-kerosine fractions for this same purpose. Many contained paraffin wax, their specific gravity lay between 0.830 and 0.900. Some residues of mixed-base oils were sold as "lubricating grease" to be used for lubricating heavy machinery, sometimes mixed with animal fats or palm oil. In the United States the Pennsylvanian refiners produced a few qualities of distilled luboils or sold the residue as such but they did not pay much attention to the quality of their products in the early days. We quote here Gale's opinion on lubricants made from petroleum:

"Rock oil is a good lubricator. In no way can this assertion be proved better than by introducing the testimony of those who by using it, have qualified themselves

[1] Thomas A. Gale, Rock Oil in Pennsylvania and Elsewhere (Erie, 1860, pp. 47-50).

to judge. A machinist who has been in the business 20 years remarks respecting the refined, that it is the best thing he can get. The owner of a factory who has applied it on light machinery for a year past, pronounces it the best thing for the purpose he ever found. He used the crude. An eminent machinist and patentee writes as follows: "Rock Oil in its crude state is a good lubricator for machinery, especially for that which has not a rapid motion. A portion of the purified article is still better. The reason being that it does not inspissate or gum."

An experienced workman states the case thus: "It is not heating, if all the old oil is cleaned off before it is applied. I never could use it on a buzz saw, till I thoroughly cleansed off the fish oil. Where that has been used, rock oil will work in with it and be worse than fish oil. By this I was deceived for several years respecting its value. That oil does not operate right with Rock oil; but tallow and this work well together. A smooth surface moving with another, plays easy and cool in it; and an arbor and mandril it will cleanse and cool. Put a little on an oil-stone and notice how clear open it makes the stone look. Should machinery stand idle six months, if it has been oiled with this, in starting, there is no clog or gum; if other lubricators have been used, there is great trouble in moving again. For cutting a screw, it appears to lack body and soon steams."

Says a gentleman largely engaged in the lumbering business: "We have three buzz, 5 feet saws on operation. We have used No. 1, Lard oil, (so marked) and Sperm No. 2, (we get no No. 1 out here) and we have used tallow, and consider rock oil the best.—We are running a shaft and pinion weighing 1000 lbs. and caps as snug as they can well be and not bind, and lubricate perhaps 4 times in 24 hours, and have no trouble with heat. All the talk about heating is because folks dont understand using it. You can't find a cooler set of running gears on the stream than these."

"Mr. F. C. Ford, Master Mechanic on the Sunbury & Erie R.R., informs us, he has used the Seneca oil on locomotives of his road and finds it admirably adapted for lubricating. The expenses on a road in the single item of oil are enormous; and if the Seneca oil possesses the qualities of a good lubricator, its cheapness must bring it into general use."

This last testimony is perhaps as important as any now cited. "Finds it admirably adapted for lubricating." A decided commendation, and it comes, as we believe the other statements do, from a competent judge—one whose reputation for care, skill and veracity would be endangered, if he did not know whereof he affirms. He has used it, and bases his statement on the trial he made. His opinion, therefore, goes far to establish the merits of Rock oil for greasing machinery, especially that of Railroads.

The following testimony is from manufactures in New Haven: "After giving your rock oil a thorough trial in our factories, we are prepared to say that as a lubricator it is superior to any oil that we have hithertho used. It is more lasting, entirely free from gum, and is not affected by heat or cold. We believe that one gallon of this oil is equal in value as a lubricator to $1\frac{1}{2}$ gallons of sperm oil.

W. & E. T. FITCH."

"I believe it can be made one of the best lubricators in use" says one who ought to know. "It improves it for car journals to put in something heavy", says another. Agreeably to this opinion, on the Cleveland & Erie R. R., Mr. Brooks, foreman, it is used mixed with lard.

"For lubricating purposes," says the Venango Spectator, "it may be used in its crude state, and is unsurpassed by any oil in the world, for heavy machinery. The residuum mixed with sperm oil has been tested for several years in a Pittsburgh cotton mill, answering the purpose better than any lubricator ever before used."

"Our steam enines at the wells use the oil just as it comes from the vats, and experienced engineers pronounce it equal to the best sperm and superior to lard oil. These engines are small, with a rapid motion, and require from their fineness of finish the best oil. It is used with satisfactory results on the steamer Venango."

With a single additional testimonial, we dismiss this branch of the subject. It comes from the enterprising proprietors of the Erie City Iron Works.—"We have used the Titusville oil in its crude state, more or less for the past year, and have put it to several tests, and we find it to have some very good qualities as a lubricator. We are now and have been for several months, using it mixed with lard oil in equal parts, and find that it makes an article superior to either alone, especially for winter use. It keeps journals clean.

„We have no hesitancy in recommending it to machinists and others running machinery, to be used as above.   LIDDELL & MARSH."

The veracity, sufficiency and force of such statements, will be admitted by most of those who have anything to do with machinery. Still, some good mechanics take a different view and condemn this new lubricator. They complain that gudgeons grind their beds, and saws and boxes heat under its application; in fine, that it lacks the body, which is vital to a good article for this purpose. And what complicates the difficulty, some express this opinion after having tried the oil, to some extent.

Touching this matter, we may say, all the localities do not produce a material dense enough in its native state. From such varieties the lighter portion must be taken off, and only the residuum employed in lubrication. In other places, as in Ohio and in Canada, the oil as it comes from the well, is thicker and heavier—more suitable to the purpose under consideration.

We respectfully suggest to all these mechanics who are incredulous in regard to the lubricating virtues of rock oil, to make a fair trial in the light supplied by the foregoing statements of their fellow craftsmen. Did our space permit, we might add, almost indefinitely, to these testimonials, others of similar purport. Can so many intelligent and disinterested judges be in error, on the subject? Make another thorough trial of the oil, mixing it as directed, if it is of a lighter grade, and see if it does not give entire satisfaction to you, as it has done to so many of your co-operators in mechanism.'"

In Europe things were very different. There a host of well-known scientists had studied the problem of friction in the hope of finding its laws and of establishing the correct properties of a good lubricant. Table II, adopted from the very valuable compilation by Vogelpohl [1] shows how famous men like Leonardo da Vinci, Newton, Amontons, de la Hire, Parent, Leibniz, Euler, von Segner and Lagrange had contributed to this growing knowledge.

[1] G. Vogelpohl, Die geschichtliche Entwicklung unseres Wissens über Reibung und Schmierung (Oel und Kohle, vol. 36, 1940, pp. 89-93; 129-134).
T. W. Chalmers, Historic Researches (London, 1950, pp. 3-24).

Ch. A. Coulomb (1736-1806) had conducted many experiments on friction without quite realizing how many factors play their part in this complicated phenomenon. He had laid down that friction was proportional to the vertical load, independent of the area of contact. It increased with the duration of undisturbed contact, grew less for wood moving over wood, and increased with the velocity of wood moving over metal. The friction was supposed to be independent for metal moving over metal. Such "laws" he deducted from the theory that small irregularities existed on the surfaces of the wood or metal he studied, which irregularities could or could not be deformed during friction. He studied the increase in viscosity rather than the lowering of the friction which follows from a closer study of the molecular structure of the

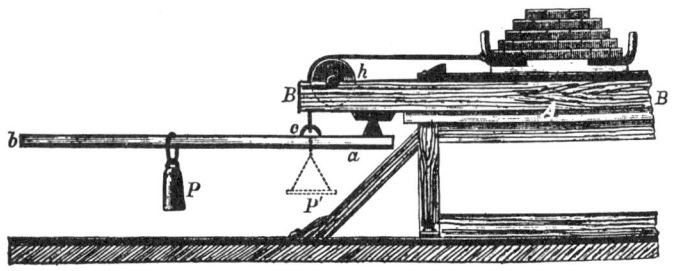

Fig. 20. Arrangement used by Coulomb for his study of sliding friction.

lubricating film on the surface of the bearings, etc. but up to the discovery of the hydrodynamic pressure in the lubricating layer (1883) such deductions from the theory of elasticity dominated the theory of lubrication. Arthur Morin (1797-1880) repeated Coulomb's experiments in the years 1831 to 1835 and though he could not find that friction increased with the duration of contact, he confirmed most of Coulomb's results. As Chalmers put it:

"For many years Coulomb's laws of friction—frequently, a little unjustly, called Morin's laws—were accepted by engineers with unquestioning assurance. They were regarded virtually as laws of nature, ranking with, and as inviolable as, the laws of motion or the law of gravitation. They certainly possessed the merit of simplicity, and were easy to apply to practical problems. Nevertheless, evidence existed from an early date which showed that at the very best they were merely convenient approximations to the truth. Morin himself had uttered a warning against taking them to be absolute and final. Subsequent research, unheeded by most engineers, soon established some of the directions in which the laws were inaccurate and incomplete. It was discovered that if the force pressing the two surfaces together was low, the resistance to sliding motion was not independent of the area of the contacting surfaces, but instead increased with the area.

These later experiments seemed to confuse rather than to elucidate existing knowledge concerning friction. The confusion was particularly marked regarding the effect of velocity. Coulomb and Morin both said that the friction of motion was independent of the velocity. Bochet's experiments implied that it decreased with the velocity. Hirn reported that it increased. A partial explanation of these discordant results might have been found in the fact that Hirn experimented with deliberately lubricated surfaces, whereas the others used ostensibly unlubricated surfaces. Such an explanation would, however, have entailed the conclusion that mediate friction was in some way or other a radically different phenomenon from immediate friction. That conclusion would have been wholly opposed to contemporary thought on the subject. It was firmly believed that mediate friction was merely an alleviated form of immediate friction, and that the use of an unguent simply reduced the value of the coefficient without altering the fundamental factors producing the friction.

Faced with this discordance of experimental evidence, engineers followed the easy course of accepting Coulomb's and Morin's findings to the complete exclusion of all subsequent contradictory results. Their textbooks encouraged them to do so. As late as 1880 they presented their readers with analytical discussions of the friction between all manner of contacting surfaces—conical and round-end pivots, footstep bearings, axles working in new and in worn bearings, and so forth—the whole of the analysis being based on the assumed universal applicability of Coulomb's laws and Morin's coefficients. To this era belongs the development of the so-called anti-friction pivot with its tractrix profile and the origination of the proposition that the friction of an axle in a new bearing is 1.57 times as great as it is when the bearing becomes worn."

Several scientists like Brix (1837) and Weisbach (1840) tried to calculate bearing-friction laws from Coulomb's results, but they did not succeed in establishing useful formulae for, as Rühlmann correctly said, they did not take into account the nature of the lubricant. Still such publications impressed many engineers with the wrong idea that the nature of the lubricant was hardly of any importance.

Fortunately a school of practical engineers set to work not reasoning from the theory of elasticity but experimenting with actual machinery. Gustave Adolf Hirn (1815-1890) discovered that as movement started the friction values fell until a certain minimum was reached after some time. This minimum value was dependent on the velocity and on the temperature of the lubricant. The temperature factor explained many of the anomalies found by earlier investigators. Hirn, who was manager of the Pechelbronn refinery, aimed at substituting mineral oils for the common fatty oils used in his own refinery, which in turn compelled him to investigate the principle of friction. This work, which, according to his notes, must have been begun in 1847, was sent to the Académie des Sciences for publication. It was refused and the secretary Fourneyron wrote that "the results were contrary to the well-known principles of mechanics", *i.e.*, the laws of Coulomb [1]. Hirn neither stopped his experiments nor his attempts to publish them and finally had his work printed [2]. He distinguished between "dry friction", in which two surfaces glide over one other, and "wet" or "indirect" friction, in which a layer of lubricant separates the two surfaces. In the latter type of friction the dominating factors proved to be pressure (load), temperature, the contact area of the two surfaces and the velocity. Hirn found much lower coefficients of friction than Coulomb and Morin. Research into the physical nature of friction led K. L. Landsberg [3] to suppose that the lubricant fills up the depressions in sliding surfaces and that real "dry friction" does not exist. He began to realize the importance of the molecular forces between the gliding surfaces and the lubricant, which were to play a large part in later investigations. Further work by Rühlmann, Conti, Poirée and Bochet let to the publications of Thurston [4] who confirmed the findings of Hirn and concluded that "the true value of a lubricant lies in its ability to reduce friction between two surfaces.

The work of the theoretical and the practical schools finally became reconciled by the efforts of Beauchamp Tower and Reynolds which Chalmers ably summarized in these words:

"Thurston's results reversed this finding, and indicated that as the velocity rose from zero the coefficient fell continuously from the static value to a minimum and then rose steadily, the minimum value being determined in any particular case by the intensity of the loading pressure. The experiments conducted jointly by Westinghouse and Galton on the friction between brake blocks and wheels showed that the co-efficient of friction decreased with the velocity and also with the intensity of the initial pressure.

The situation had now reached such a stage of confusion that in 1879 the Institution of Mechanical Engineers appoiunted a committee to study "friction at high velocities, specially with reference to friction of bearings and pivots, friction of brakes, etc.". The Committee consisted of E. A. Cowper, Captain Douglas Galton, Dr. John Hopkinson, Professor Fleeming Jenkin, John Ramsbottom, Lord Rayleigh and Professor A. B. W. Kennedy. There was some delay in starting the Committee's work, but during 1882 it succeeded in securing the services of Beauchamp Tower for the purpose of carrying out some experiments on a specially designed machine at the Edgware Road works of the Metropolitan Railway. Tower threw himself into

---

[1] Extrait des Comptes Rendus Hebdomadaires Acad. des Sciences, vol. XCIX, séance du 1er déc. 1848.

[2] A. G. Hirn, Les lois de frottement (Bull. Soc. Ind. de Mulhouse, vol. 26, 1854, pp. 128 ff.).

P. Edm. Schmitz, L'oeuvre scientifique de G. A. Hirn (Revue Technique Luxembourgeoise, vol. 50, 1958, No. 1, pp. 27-33).

[3] Ann. d. Physik vol. 121, 1864, pp. 283 ff.

[4] R. H. Thurston, Friction and Lubrication (London, 1879).

R. H. Thurston, Friction and Lost Work in Machinery, (New York, 1885).

the work with enthusiasm, and by September, 1883, submitted his first report to the Committee. The report was presented to the members at the Autumn Meeting of the Institution, held at Birmingham on November 1st of that year.

The immediate result of Tower's first series of experiments was the recognition by practical engineers of the fact that "dry" friction and "lubricated" friction were fundamentally different phenomena which had nothing in common except an unfortunate similarity of name."

A short second report was published in 1885:

"Before beginning the experiments Tower ran the machine for some time under a heavier load than that used during the tests, the object being "to wear the brass down to a perfect fit on the journal". He found that in taking off or putting on the load the oil pressure fell or rose in exact proportion to the loading. At the end of the experiments the speed was reduced from 150 r.p.m. to 20 r.p.m. Exactly the same oil pressures were recorded at the low as at the high speed.

Concerning the third (1888) and fourth (1891) reports little need be said. The third dealt with the friction of a collar on a shaft, the experimental method employed consisting of pressing two rotating discs against the faces of an interposed annulus and measuring the effort required to prevent the annulus from rotating with the discs. The fourth report related to the friction of a plane-ended footstep bearing, the procedure adopted being to support a loaded vertical shaft in a cup bearing at its lower end and to determine the effort needed to hold the cup stationary when the shaft was rotated. The machines were of complicated design and in both cases the lubrication of the contacting surfaces was affected in an arbitrary specialised manner. In spite of all the care lavished on the experiments, the results obtained were characterised by irregularity. The revealed nothing of a striking fundamental nature and are to-day of little interest.

No one studying the first and second reports can fail to be impressed by Tower's genius as an experimenter. Guided, it might almost seem, by instinct, he laid bare in a simple, direct and convincing manner hitherto unknown fundamental facts which set the stage for a new understanding of friction and lubrication. When we pass on to the third and fourth reports we become conscious of a decided change of atmosphere. Complication of method and apparatus replaces the previous simplicity. The results recorded are marked by indecision and uncertainty. They have no fundamental importance or general applicability and are almost entirely peculiar to the particular machine and method of lubrication employed.

Tower found his interpreter in Osborne Reynolds, Professor of Engineering at Owens College, Manchester. Reynolds' analysis of Tower's experimental results was presented in a lengthy paper "On the Theory of Lubrication and Its Application to Mr. Beauchamp Tower's Experiments, including an Experimental Determination of the Viscosity of Olive Oil", published in the "Philosophical Transactions" of the Royal Society, Part I, 1886. It was a very remarkable document. It presented a complete mathematical study of the hydrodynamical action taking place in a lubricated journal bearing. The formulae arrived at predicted the manner in which the resistance should vary with the load and the velocity, the pressure which should exist in the oil film between the brass and the bearing, and the variation of that pressure from point to point of the surfaces. On the insertion of the data appropriate to Tower's experiments the formulae yielded numerical values in almost exact accord with the experimental values given in Tower's reports.

Accepting these modern views, we can readily understand the fundamental physical difference between the friction of lubricated surfaces as studied by Tower and the friction of dry or nominally dry surfaces. The one is wholly determined by the viscosity of the lubricant. The other is wholly or in part determined by the capacity of the metals constituting the two surfaces to seize together and the restance of the welded bridges to rupture by shearing."

If we now turn to the production of lubricants in Europe we find that petroleum lubricants had to fight for their position amongst a market crowded with various competitors. Animal and vegetable oils like sperm oil, olive oil and rapeseed oil were still in common use. Lubricants were also manufactured from the shales of Autum and other localities in France, from lignites of Germany by Messrs Wiesmann & Co. (Bonn), Baumeister & Co. (Bitterfeld), B. Hübner (Rehnsdorf), Rolle & Co. (Saxony and Thuringia), from Galician crude and Ozokerite from Moldavia by Wagemann & Co. (Vienna), from "Rangoon tar" by Field and Price (Belmont), from "Pechelbronn oil sands" by Hirn and from cannel-coal in Scotland. Some even used their own mixtures and recipes such as two parts of ordinary petroleum and one part of terebinth oil [1]. Ready-made luboils were imported from Russia, such as Petroffsky's Kaukasine and the Oleonaphtha of Ragosine and Comp. [2].

Again the European market was soon swamped with cheap American luboils such as Globoil, Topazoil, Staroil, Valvoline, Rubinoil and Vulcanoil. Most of these were residual oils decolorized with carbon-black but some were badly fractionated distillate. The Vulcanoil of the Vulcanic Oil and Coal Company of West Virginia had been lavishly advertized at the Paris Exhibition of 1867 and several laudatory articles appeared in the European press [3] claiming that it dissolved grit and dirt from the bearings, had a flashpoint so high "as to constitute no danger to the surroundings" and was cheap in practice as "it could not be stolen and burnt in lamps like rapeseed oil". It was sold in drums of 280 lbs. net.

However, European consumers were very wary and though they were convinced in certain respects such as the much slower rate of oxidation of mineral luboils and their slight tendency to emulsification with condensed water, they found that resinous products were formed in the long run and

---

[1] L. Bechstein, Le pétrole employé dans le travail au tour des métaux et alliages très durs (Le Technologiste vol. XXX, 1868, p. 390).

[2] Anon., Sur les huiles minérales employées au graissage des véhicules et des machines (Le Technologiste vol. XXX, 1869, pp. 41-47).

[3] R. Zenker, Mineralöle zum Maschinenschmieren (Dtsche Ind. Ztg. 1867, p. 417). Kayser, Kesseler und Schmelzer. Die Anwendung vom Mineralölen zum Schmieren vom Maschinen (Dtsche Ind. Ztg. 1867, p. 396).

that they were unsuitable as steam cylinder oils because of their low flash point. Walkhoff used them with some success in the textile industry but he found that lignite luboils, *e.g.*, those made by Rolle & Co., were more viscous and stable.

Farez and Boulanger [1] laid down that a good lubricant should have a sufficient "unctuosity" (viscosity) to maintain a lubricating layer under service conditions, and sufficient stability ("inalterability"), in which mineral oils

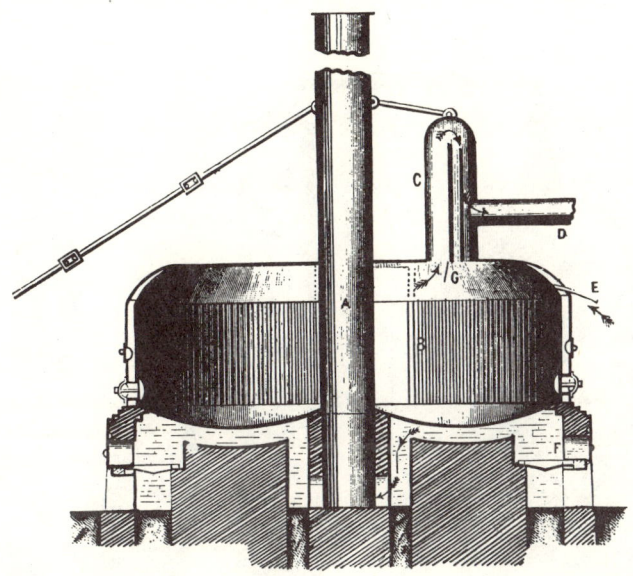

Fig. 21. "Cheese-box still" used around 1860 for topping crude and for the distillation of lubricating oils.

excelled. The latter also possessed the required "neutrality" towards the metal surfaces to be lubricated, for the acid parts of vegetable and animal oils had always to be eliminated first. The American luboils introduced into Europe during the last 15 years, they said, were mostly badly distilled and cracked, hence their low flash point, which was sometimes not more than 60-65°C. Steam distillation would have cured this immediately. Heavy European luboils over 0.860 had flash points of 135-145°C but the American residual oils had barely 100°C and when cooled formed an amorphous mass. Still when properly distilled Pennsylvanian crude gave excellent luboils, but the steam used had

[1] Farez et Boulanger, Mémoire sur les huiles employées au graissage des machines (Bull. Soc. Industr. Mulhouse vol. XLV, 1875, pp. 302-325).

to have a temperature of at least 220°C. Such luboils were found to form "resinous products" only after one to two years exposure to the air and they did not corrode metal surfaces as experiments with copper, bronze and steel plaques proved. Farez tested such luboils in spinning machines at Douai, Cherbourg and Hamégicourt (Aisne) with great success and since then some hundreds of tons of mineral luboils have been used in the textile industry there.

The European refiner was not blessed with an excellent base material for luboil manufacture such as Nature had provided for his Pennsylvanian col-

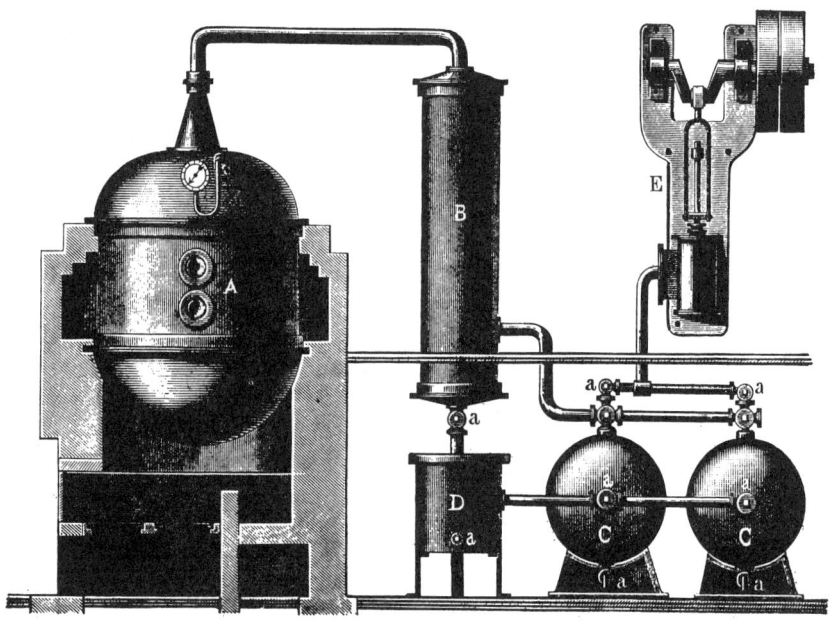

Fig. 22. The vacuum distillation of lubricating oils about 1875.

league. He had crudes of very different origin at his disposal from which he had to manufacture luboils to compete with others from different industries and he had to devise refining methods to prepare a more or less standard product from these various crudes. Sometimes he had to spend much ingenuity to eliminate "asphaltic compounds" and paraffin wax from his distillates or residual oils. Hence he devised tests to determine the properties of his intermediary and refined products. The principal property of the lubricant was its viscosity. This was still tested in some very primitive apparatus such as that of Nasmyth [1], which the author described in this way:

"The most valuable quality in an oil intended for the lubrication of machinery is permanent fluidity. That oil which will for the greatest length of time remain fluid in contact with the iron or brass is, without doubt, the most useful for the purpose. Hence, as before said, the necessity of including the element of time in any experiment on the comparative value of such oils.

Some idea may be formed of the importance of having the means of arriving at correct conclusions on this subject, when we know that in some spinning establishments there are upwards of 50,000 spindles in motion at the rate of 4,000 or 5,000 revolutions per minute! The slightest defect in the quality of the oil in such a case, by its becoming viscid, tells in the most serious way upon the quantity of fuel consumed in generating the power required to maintain at this high velocity such a multitude of moving parts. The slight increase of fluidity consequent on the rise of temperature, caused by the lighting of the gas in the rooms of a cotton-mill, makes a difference of several horsepower in the duty of the engine of an extensive establishment.

The oil test we have now to describe, and which is an invention of Mr. Nasmyth's, consists of a plate of iron 4 inches wide by 6 feet long, on the upper surface of which six equal-sized grooves are planed. This plate is placed in an inclining position, say 1 inch in 6 feet. The mode of using it is as follows: — Suppose we have six varieties of oil to test, and we are desirous to know which of them will for the longest time retain its fluidity when in contact with iron and exposed to the action of the air; all we have to do is to pour out simultaneously at the upper end of each inclined groove an equal quantity of each of the oils under examination. This is very conveniently and correctly done by means of a row of small brass tubes. The six oils then make a fair start on their race down hill; some get a head the first day, and some keep a head the second and third day, but on the fourth or fifth day the truth begins to come out; the bad oils, whatever good progress they may have made at the outset, come soon to a standstill by their gradual coagulation, while the good oil holds on its course, and at the end of eight or ten days there is no doubt left as to which is the best; it speaks for itself, having distanced its competitors by a long way. Linseed oil, which makes capital progress the first day, is set fast after having travelled 18 inches, while second-class sperm beats first-class sperm by 14 inches in nine days, having traversed in that time 5 feet 8 inches down the hill."

Nasmyth's apparatus was improved by Bailey (1878) but later refiners rightly preferred the viscometer of the type we still use [2]. Charles Dolfuss addressed the Société Industrielle de Mulhouse on June 29, 1831, demonstrating an instrument for the testing of lubricants, a vessel with a small hole in the bottom, from which the lubricant could escape. By measuring the

---

[1] Mr. Nasmyth's Test for Oils for Lubricating (J. Franklin Institute vol. L, 1850, pp. 403-404).

[2] F. Fischer, Über die Untersuchung von Schmierölen (Dingler's Polyt. J. vol. 236, 1880, pp. 487-497).

number of seconds necessary for a measured volume of lubricants to leave the vessel he obtained an index for its liquidity. This instrument Dolfuss dubbed a "viscomètre". This viscometer was improved by Vogel (1863) and Coleman (1873). Fischer employed a platinum capillary surrounded by an outer waterbath to keep its temperature constant and to heat the luboil in the inner vessel, the temperature of which was carefully measured. This was indeed already a close approach to the later Redwood or Engler apparatus.

Fischer also states that mineral luboils are often mixed with vegetable oils tallow, etc. By alcoholic saponification the fatty acids can be separated and identified and colour and smell of the lubricant should be noted too. The specific gravity proved a good test for contaminants. The melting point of greases and the solidification point of oils should be determined. Resins can be separated with alcohol; free acids are measured by volumetric analysis. The value of a lubricant can also be tested in such machines as Hirn and Thurston had developed but one should be careful how to interpret the results. We reproduce a table compiled by Fischer giving some properties of crudes, residues and luboils which were marketed in his days. He concluded that paraffin wax was a bad lubricant which had too little viscosity at high temperatures.

Thurston's book presents the accumulated knowledge on lubricants by 1880 in a well-balanced form. He admires sperm oil as a lubricant but this is far too expensive for common use. Most vegetable oils contain too much acid or form gum readily. Hence, though there was a range of liquid, semi-solid (tallow, soap and wax) or solid (graphite, soapstone) lubricants one had to make one's choice carefully to pick the appropriate lubricant for the specific case, keeping in mind the type of apparatus to be lubricated and above all the bearing-metal. In his own words [1]:

"The value of a lubricant to the consumer, as is seen from what has been just stated, depends on its cost in the market, its efficiency in reducing friction, its durability under wear, its freedom from liability to "gum", its freedom from acid and from grit, and its permanence of composition and of physical condition when subjected to changes of temperature, and also, frequently, its capacity for carrying away heat from journals already heated.

The most generally applied fluid lubricants are the better known and cheaper kinds of animal, fish, vegetable and mineral oils; of these sperm stands admittedly at the head of the list, lard, neats-foot, whale, tallow, seal and horse oils are all largely used either alone or mixed. The vegetable oils in use are olive, which is by far most generally used in other countries; cotton-seed oil in the United States, palm, rape-seed, oleine, colza, poppy, peanut, rosin, cocoanut and castor oils are all more or less employed in lubrication. Of the fish oils, porpoise, black-fish, cod and

[1] R. H. Thurston, Friction and Lubrication (London, 1879, pp. 45-55, 211-212).

	original colour	specific gravity at 15°	shaken up with water	shaken with caustic potash solution	shaken with sulphuric acid	nitric acid temperature increase	discharge time at 10°	discharge time at 40°
Water	—	1.000	colourless	—	—	—	38 secs	25 secs
Rock oil from Steinwörde	dark brown	0.9401	colourless	yellow	yellow	5.4°	indeterminable	1 465
Rock oil from Wietze	dark brown	0.9460	slightly cloudy	yellow	yellow	8.2	11 450	1 175
Rock oil from Hölle bei Heide	dark brown	0.9395	colourless, clear	colourless	yellowish cloudy	12.1	4 380	596
Rock oil from Odessa	brown	0.9089	colourless, clear	colourless	almost colourless	4.8	1 335	202
Rock oil from Sehnde	dark brown	0.8498	colourless, clear	colourless	yellow	10.9	79	43
Finest spindle oil	slightly yellow	0.8700	colourless, clear	colourless	slightly yellow	0.7	465	118
Amber spindle oil	orange-yellow fluorescing greenish	0.8698	colourless, clear	colourless	yellowish, oil darker	1.3	472	120
Finest machine oil	brown, fluorescing green-blue	0.8797	colourless, clear	colourless	yellow, oil darker	2.7	2 695	368
Cylinder oil	brown, fluorescing green-blue	0.8904	colourless, clear	colourless	yellow, oil darker	2.6	12 060	1 090
Lignite-tar oil	brown	0.8911	colourless, clear	yellow, oil darker	black-brown	45.6	71	42
Colza oil	light brown yellow	0.9169	slightly cloudy	soapy	greenish, oil	2.5	615	191
Olive oil (wood oil)	bright yellow	0.9178	colourless	soapy	green-blue yellowish	1.6	605	168

"Valvolines" brackets: Finest spindle oil, Amber spindle oil, Finest machine oil, Cylinder oil menhaden oils are most used. The mineral oils are of two general classes; the shale oils, obtained from certain shales, and the petroleums, which come from extensive oil lakes, situated, usually, far beneath the surface of the earth, and which are principally supplied by the oil wells of Pennsylvania and other of the United States.

Of these oils, sperm is still largely used, notwithstanding its high price, since it excels nearly all others in its power of reducing friction and immensely excels them in endurance. Rape-seed is, in some districts, now displacing olive oil as a lubricant, but the mineral oils, pure or mixed, are rapidly taking the leading place in all markets. Scotland is producing 25,000,000 gallons of shale oil from 800,000 tons of shale, of which oil 10,000,000 gallons are illuminating oils and the remaining products include 10,000 tons of lubricating oil, 6,000 tons of parafine wax, and also 2,500 tons of ammonium sulphate. The petroleums are found in China, India, Italy and other parts of the world; the island of Trinidad contains a lake of petroleum—,,Pitch Lake"—the shores of which are composed of bitumen, produced by its evaporation and oxidation. But the greater part of the petroleum of the world is produced in Pennsylvania, West Virginia and Ohio. Pennsylvania alone has produced in a single year—1874—over 10,000,000 barrels, of which one-half was exported to Europe and used largely for lubrication, but far more generally in illumination.

The greases, or semi-fluid lubricants, are sometimes used in their natural state, as tallow, lard, wax and other similar substances and sometimes are made up artificially, *e.g.*, the various kinds of soap.

Mixtures of tallow and black-lead, white-lead and oil, and other mixtures containing sulphur, are often used. For some special purposes, certain mixtures are used, as, for cooling hot journals, mixtures of oil and of white or black lead, oil and sulphur or greases composed of oil to which some alkaline water has been added. I have, in my experience with very large and troublesome marine engines, found sulphur and oil on the journal, with the application—very cautiously—of cold water externally, to work best.

The only method of learning the true value of a lubricant, and its applicability in the arts, is to place it under test, determining its friction-reducing power, and its other valuable qualities, not only at a standard pressure and velocity, and at ordinary temperatures, but measuring its friction and endurance as affected by changing temperatures, speeds and pressures throughout the whole range of usual practice.

The true value of an oil to the consumer is not proportional simply to its friction-reducing power and endurance under the conditions of his work; but its value to him is measured by the difference in value of power expended, using different lubricants, less the difference in total cost of oil or grease used; but for commercial purposes no better method of grading prices seems practicable than that here adopted, which makes their market value proportional to their endurance, divided by their coëfficients of friction. The consumer will usually find it economical to use that lubricant which is shown to be the best for his special case, without regard to price, and will often find real economy in using the better material sufficient to repay excess in the total cost very many times over. He cannot afford to accept low grade unguents, even without charge."

3. *Bitumen for Roofs, Shingles and Pipes*

In the days of Drake natural asphalt and bituminous mastic had already become a standard material in all roadbuilding handbooks [1]. The residues of petroleum refining could not only be turned into good lubricants and sold as such or as liquid fuel, it could be inspissated to asphaltic bitumens of various properties for which new uses had been found. Though few residues were turned into asphaltic bitumens before 1880 (and indeed before the crudes of Venezuela and Mexico provided the industry with suitable base materials in lavish quantities about 1910) some interesting applications had been discovered. For most of these mastics prepared from natural asphalt or other bituminous residues such as coal-tar pitch were used. This is what Ure [2] has to say about the properties of such mastics:

"It is a very remarkable fact, in the history of the useful arts, that asphalt, which was so generally employed as a solid and durable cement in the earliest constructions upon record, as in the walls of Babylon, should for so many thousand years have fallen well nigh into disuse among civilised nations. For there is certainly no class of mineral substances so well fitted as the bituminous, by their plasticity, fusibility, tenacity, adhesiveness to surfaces, impenetrability by water, and unchangeableness in the atmosphere, to enter into the composition of terraces, foot-pavements, roofs, and every kind of hydraulic work. Bitumen, combined with calcareous earth, forms a compact semi-elastic solid which is not liable to suffer injury by the greatest alternations of frost and thaw, which often disintegrate in a few years the hardest stone, nor can it be ground to dust and worn away by the attrition of the feet of men and animals, as sandstone, flags, and even blocks of granite are. An asphalt pavement, rightly tempered in tenacity, solidity, and elasticity, seems to be incapable of suffering abrasion in the most crowded thoroughfares; a fact exemplified of late in a few places in London, but much more extensively, and for a much longer time, in Paris.

Numerous experiments and observations have led me to conclude that fossil bitumen possesses far more valuable properties for making a durable mastic than the solid pitch obtained by boiling wood or coal tar. The latter, when inspissated to a proper degree of hardness, becomes brittle, and may be readily crushed into powder; while the former, in like circumstances, retains sufficient tenacity to resist abrasion. Factitious tar and pitch being generated by the force of fire, seem to have a propensity to decompose by the joint agency of water and air, whereas mineral pitch has been known to remain for ages without alteration.

Bitumen alone is not so well adapted for making a substantial mastic as the native compound of bitumen and calcareous earth, which has been properly called asphaltic rock, of which the richest and most extensive mine is unquestionably that of the Val de Travers, in the canton of Neufchâtel. This interesting mineral deposit occurs in the Jurassic limestone formation, the equivalent of the English oolite. The mine is very accessible, and may be readily excavated by blasting with gunpowder. The stone is massive, of irregular fracture, of a liver-brown colour, and is interspersed with a few minute spangles of calcareous spar. Though it may be scratched with the nail, it is difficult to break by the hammer. When exposed to a very moderate

---

[1] R. J. Forbes, Studies in Early Petroleum History vol. I (Leiden, 1958, Chap. II).
[2] Ure's Dictionary of Arts, Manufactures and Mines (London, 1860, Vol. I, pp. 310-313).

heat it exhales a fragrant ambrosial smell, a property which at once distinguishes it from all compounds of factitious bitumen. Its specific gravity is 2.114, water being 1.000, being nearly the density of bricks. It may be most conveniently analysed by digesting it in successive portions of hot oil of turpentine; whereby it affords 80 parts of a white pulverulent carbonate of lime, and 20 parts of bitumen in 100. The asphalt rock of Val de Travers seems therefore to be far richer than that of Pyrimont, which, according to the staetment in the specification of Claridge's patent of November, 1837, contains "carbonate of lime and bitumen in about the proportion of 90 parts of carbonate of lime to about 10 parts of bitumen".

The calcareous matter is so intimately combined and penetrated with the bitumen as to resist the action not only of air and water for any length of time, but even of muriatic acid; a circumstance partly due to the total absence of moisture in the mineral, but chiefly to the vast incumbent pressure under which the two materials have been incorporated in the bowels of the earth. It would indeed be a difficult matter to combine, by artificial methods, calcareous earth thus intimately with bitumen, and for this reason the mastics made in this way are found to be much more perishable. Many of the factitious asphalt cements contain a considerable quantity of siliceous sand, from which they derive the property of cracking and crumbling down when trodden upon. In fact, there seems to be so little attraction between siliceous matter and bitumen, that their parts separate from each other by a very small disruptive force.

Since the asphalt rock of Val de Travers is naturally rich enough in concrete bitumen, it may be converted into a plastic workable mastic of excellent quality for foot pavements and hydraulic works at very little expense, merely by the addition of a very small quantity of mineral or coal tar, amounting to not more than 6 or 8 per cent. The union between these materials may be effected in an iron cauldron, by the application of a very moderate heat, as the asphalt bitumen readily coalesces with the tar into a tenacious solid.

In the able report of the Bastenne and Gaujac Bitumen Company, drawn up by Messrs. Goldsmid and Russel, these gentlemen have made an interesting comparison between the properties of mineral tar and vegetable tar: the bitumen composed of the latter substance, including various modifications extracted from coal and gas, have, so far as they were able to ascertain, entirely failed. This bitumen, owing to the qualities and defects of vegetable tar, becomes soft at 115° of Fahrenheit's scale, and is brittle at the freezing point; while the bitumen into which mineral enters will sustain 170° of heat without injury. In the course of the winter, 1837-38, when the cold was at $14\frac{1}{2}$° below zero C., the bitumen of Bastenne and Gaujac, with which one side of the Pont Neuf at Paris is paved, was not at all impaired, and would, apparently, have resisted any degree of cold; while that in some part of the Boulevard, which was composed of vegetable tar, cracked and opened in white fissures. The French Government, instructed by these experiments, has required, when any of the vegetable bitumens are laid, that the pavement should be an inch and a quarter thick; whereas, where the bitumen composed of mineral tar is used, a thickness of three-quartes of an inch is deemed sufficient. The pavement of the bonding warehouses at Bordeaux has been laid upwards of 15 years by the Bastenne Company, and is now in a condition as perfect as when first formed. The reservoirs constructed to contain the waters of the Seine, at Batignolles, near Paris, have been mounted 6 years, and notwithstanding the intense cold of the winter of 1837, which

NEW BYPRODUCTS FOR BURNERS, BEARINGS AND BITUMENS

froze the whole of their contents into one solid mass, and the perpetual water pressure to which they are exposed, they have not betrayed the slightest imperfection in any point. The repairs done to the ancient fortifications at Bayonne, have answered so well, that the Government many years ago entered into a very large contract with the company for additional works, while the whole of the arches of St. Germain and St. Cloud railways, and the pavements and floorings necessary for these works, have been laid with Bastenne bitumen.

The mineral tar in the mines of Bastenne and Gaujac, is easily separated from the earthy matter with which it is naturally mixed, by the process of boiling, and is then transported in barrels to Paris or London, being laid down in the latter place to the company at 17 £. per ton, in virtue of a monopoly of the article purchased by the company at a sum, it is said, of 8000 £.

Mr. Harvey, the superintendent of the Bastenne Company, was good enough to supply me with various samples of mineral tar, bitumen, and asphaltic rock for analysis. The tar of Bastenne is an exceedingly viscid mass, without any earthy impurity. It has the consistence of bakers' dough at 60° of Fahrenheit; at 80° it yields to the slightest pressure of the finger; at 150° it resembles a soft extract; and at 212° it has the fluidity of molasses. It is admirably adapted to give plasticity to the calcareous asphalts.

A specimen of Egyptian asphalt which he brought me, gave, by analysis, the very same composition as the Val de Travers, namely, 80 per cent. of pure carbonate of lime, and 20 of bitumen. A specimen of mastic prepared in France was found to consist, in 100 parts, of 29 of bitumen, 52 of carbonate of lime, and 19 of siliceous sand. A portion of stone called the natural Bastenne rock afforded me 80 parts of gritty siliceous matter and 20 of thick tar. The Trinidad bitumen contains a considerable portion of foreign earthy matter; one specimen yielded me 25 per cent. of siliceous sand; a second, 28; a third, 20; and a fourth, 30: the remainder was pure pitch. One specimen of Egyptian bitumen, specific gravity 1.2, was found to be perfectly pure, for it dissolved in oil of turpentine without leaving any appreciable residuum. As the specific gravity of properly made mastic is nearly double that of water, a cubic foot of it will weigh from 125 to 130 lbs.; and a square foot, three quarters of an inch thick, will weigh very nearly 8 pounds. A ton of it therefore will cover 280 square feet. The prices at which the Bastenne Bitumen Company sell their products are as follows:

Pure Mineral tar, 24 £. per ton, or 28 s. per cwt.
Mastic, 8 £. 8 s. per ton, or 10 s. per cwt.

			Side Pavement		Roofs and Terraces
From	50 to	100 feet,	1 s.	3d. per foot	1 s. 6d. per foot
	100	250	1 s.	1d.	1 s. 4d.
	250	500		11d.	1 s. 1d.
	500	750		10d.	1 s. 0d.
	750	1000		9d.	11d.
	1000	2000		8d.	10d.
	1000	2000		8d.	10d.
	2000	5000		7d.	9d.

Where the work exceeds 5000 feet, contracts may be entered into. For filling up joints of brickwork, etc. 1d. to 1¼d. per foot run, according to quantity. These prices are calculated for half an inch thickness, at which rate a ton will cover 420 square feet.

As the Val de Travers Company engage to lay down their rich asphaltic rock in London at 5£. per ton, and as a mineral tar equal to that of Seissel may probably be had in England at one fourth of the price of that foreign article, they may afford to lay their mastic three quarters of an inch thick per the thousand feet, including a substratum of concrete, at a rate of fivepence a square foot instead of fifteenpence, being the rate charged under that condition by the Bastenne Company."

The inventor of the shingle is a Dr Arvid Faxe (or Faxa), a councellor and physician to the Swedish Admiralty, who on March 7, 1787 read a paper to the Swedish Academy on the matter of a new method of roofing [1]. He used paper or cardboard shingles impregnated with copperas and nailed to the planking of the wooden roof and then painted them with hot wood-tar to waterproof them. They did not prove very successful as they weathered quickly, but in the same year Oberbaurath Gilly introduced them in Germany and discussed their merits in his handbook on architecture. Notices on Faxe's invention appeared in certain foreign periodicals [2] and we have detailed descriptions of their composition [3]. There are also two long articles on their application in Germany [4] which we give here in English translation:

"*Announcement of a German Factory for Faxe's Pasteboard for the Protection of Woodwork from Fire.*

Some years ago a certain Doctor Arfrid Faxe of Carlscrona announced in various publications that he had produced a so-called fire-proof pasteboard or roofing-paper which, when spread over the woodwork and nailed down, not only protects it against attack by the most violent fires but also, as a covering on roofing boards, withstands unchanged frost, damp, heat and all changes in the weather just as well as a covering of tiles or lead. The fire tests made on this pasteboard in Sweden, Berlin and several other places have been completely satisfactory, but hitherto it has not been possible to make any conclusive tests to ascertain the qualities of this pasteboard as roofing. It is not surprising that the public should distrust all proposals and inventions for the fireproofing of woodwork since so many proposals and inventions of this kind have been published for more than 20 years which lacked the essential

---

[1] Dr. Arvid Faxe, Inträdes-Tal on Sten-Papper Hallet (Stockholm, Lange, 1787).

[2] Göttinger Anzeiger von gelehrten Sachen vol. III, 1787, pp. 1940-1943; Journal des Savants 1788, No. 6, p. 994.

[3] Magazin f. d. Neueste der Physik und Naturgeschichte vol. IV, 1786/87, part i, p. 176 part 4, p. 40.

[4] C. Luhmann, Die Fabrikation der Dachpappe und die Anstrichmasse für Pappdächer (1e edit. Hartleben, Vienna, 1883); (2 edit. Hartleben, Wien, 1902).
Anon., Über die Herstellung von Dachpappe in Schlesien in den Jahren 1787 und 1793 (Teer und Bitumen, vol. 33, 1935, pp. 365-367).

requirements of reliability, durability, practicability and economy. Hence the person who under these circumstances once more attempts to publicise a means of protecting woodwork against the effects of fire and weather and to recommend it as an unfailing method, combining all the essential qualities which were lacking in the previous suggestions and inventions, will have to be all the more careful and sure of himself. If, however, the value of the truth can be increased and made public by tests which are reliable and easy to carry out, there is still hope that the public's hopes, which have so often been deceived by newly invented fire- and water-proof coverings, will be realised.

The said roofing is neither an ideal nor a fraud, but a real thing which will certainly commend itself to anyone using it according to the instructions. In recommending this invention we should not be afraid of intelligent examiners who have sensible misgivings, although it would be desirable if hasty investigators with foolish prejudices were to reserve judgement until the former have shown the way with cold, unbiassed conviction based on experience, after which the latter will be at liberty to echo them. The objections to the fact that the roofing made in Carlscrona is a foreign product and that money paid for it will go out of the country, that the inventor makes a secret of its manufacture, that it is even extraordinarily difficult to obtain samples of the pasteboard from there, and that the costs would outweigh the advantages, will all be overridden at once when the public is reliably informed that the Provincial Treasurer at Breslau, Mr. Herzberg, has made roofing after the style of samples from Carlscrona in such a way that no difference in appearance or effect can be seen between the original and the copy. He has also carried out a convincing experiment to show that this pasteboard may be reliably used as a covering on boarded roofs, for he covered a flat boarded roof with it last autumn which has withstood damp, frost and heat up to now without being affected and in respect of durability it promises to be as serviceable as a lead covering.

In pursuance of the public announcement Dr. Faxe of Carlscrona has been given both the sole right to manufacture the pasteboard and also received 300 silver Reichsthaler from the royal privy purse for the purpose of extending and improving the factory. This proves that Sweden rewards industry and protects inventions which are useful to all. Silesia can also be proud of the fact that no single matter directly or indirectly bearing on the common good escapes the ever-watchful eye of the Excellency Count von Hoym, the chief Royal Minister of Finance and War, and that he is unceasingly engaged on the promotion of undertakings aimed essentially at the general good and on encouraging them by his active support. His Excellency has not only instructed Mr. Herzberg to perfect the copied pasteboard, but in order to encourage and help this gentleman to establish and equip a pasteboard factory has graciously granted him a premium of 300 Reichsthaler.

The erection of such a factory is already under way and the public may confidently hope to obtain in future years all kinds of pasteboard so as to carry out tests with it and convince themselves of the truth, and then to use this decidedly useful product on a large scale for the purpose intended. The directions for the various uses of the pasteboard will then be published in detail. Although this pasteboard will be its own recommendation when properly used, it will not be quite beside the point to state what great progress has already been made in its use in Sweden.

That it withstands fire has been adequately established, and that the pasteboard imitated by Mr. Herzberg could also be quite safely used as a roof covering is shown by his above-mentioned experiment with the boarded roof of which everyone can now convince himself by looking at it. As a sheet of pasteboard the size of a sheet of cancelli paper, if the fabric is in fashion and has a market, with not more cost than 9 pfenning, this means that such a roof covering is not only extremely cheap, but requires at least 1/3 less timber in the shape and construction of the roof-work itself, because roofs covered with stone slabs can be laid much lower and flatter than usual.

The following are not specially included in the main advantages of using this pasteboard which have already been proved by experience in Sweden.

1. When the ceilings of a solid house are covered with it instead of rushes they will not burn through, but persons living in the upper rooms can remain quite calm during a fire in the lower rooms. To achieve the same end with wooden walls they are covered with sheets of this pasteboard.

2. It can also be used in gutters, and the bottoms of sea-going vessels may be protected from attack by worms by coating with this material; convincing tests have been made on this in Sweden.

3. Busts and vases can also be made from the paste of this material which withstands all kinds of weather.

It will perhaps be sufficient for the present to draw the public's attention to a material the value of which speaks for itself and makes any further recommendations unnecessary.

Pasteboard for the Protection of Woodwork against Fire.

The German factory of Faxe's pasteboard for the protection of woodwork against fire, which was announced on page 147 of volume six of the Schlesische Provinzialblätter, has unfortunately not been established owing to the death shortly afterwards of the builder, the Provincial Treasurer in Breslau, Mr. Herzberg. After his death a certain Johann Drescher of Breslau, a former householder in the capital province made similar attempts. As he informed the Royal Breslau Ministry of War and Domains, he had constantly assisted Mr. Herzberg in this work and became so well-informed of the whole secret and all its advantages that he had sufficient knowledge to build a factory of that kind. He only needed support. To obtain it he offered to give a small proof of the genuineness of the pasteboard made by him. As the Royal Chamber thought the matter was even worthy of attention, it instructed the magistrate at Breslau to carry out this test. It was performed on 5th April last by councillor Witte in the presence of the town master-builder Mr. Dreyer, the master carpenter Mr. Krause, and the chimneysweep Mr. Graeser. Mr. Drescher provided for the test a model of a house about 1 ell in height up to the ridge beams, 1 ell long and $\frac{1}{2}$ an ell broad covered on top and round about with paste. It was heaped with straw, twigs and small pieces of wood. When this material had been set alight and burned it was seen that the paste had not burned but had become very brittle so that it easily peeled off and this was even more the case after repeated attempts and moistening with warm water. As the fire which had been lit had not been strong enough to make the paste red hot, the experts present ordered another test to be made, since, if the paste were to absorb this degree of heat its advantage would be greatly reduced because in a big fire it would become red hot and could set the rafters alight. This test was carried out and the paste became red hot, but the timber underneath was only singed and did not catch fire. In the experts opinion this paste is preferable to shingle, straw or rush roofing, because it entirely prevents fire caused by sparks and does not distintegrate in fierce heat, and even when it is red hot the timber underneath it does not catch fire, but cools again immediately. Furthermore, according to Drescher's statement, such a roof must be trimmed and cemented, thus making it impossible for a fire to reach the rafters.

According to Drescher's statement, his paste also has the advantages of being light and cheap. Four sheets, covering an area of 40 pieces of planking only weigh as much as one piece of planking. Mr. Drescher has already promised for 3 or 4 shillings a sheet for an area which would require 10 roof tiles which would be the cost of the tiles without counting the cost of nails and laths, and if large-scale manufacture is undertaken in the future to supply them so cheaply that the roofing will not cost more than straw or rushes. It appears then that this roofing could be advantageous in the country on buildings which for reasons of economy do not need a warmer roof, and could at any rate be useful in that they will not be softened and broken up by weathering and the passage of time. Should a test be made of this, the result will be reported in due course as will the further progress of this undertaking."

These developments were arrested by the Napoleonic Wars but experiments were revived by 1820 when Michael Kag of Mühldorf (Bavaria) announced [1] that raw paper saturated with varnish and coated with a mineral powder would make an excellent roofing and could also be recommended as a substitute for leather for the soles of shoes.

Again the Magdeburger Zeitung of November 16, 1822 contained a notice on paper impregnated with tar to be used instead of straw and wooden shingles on roofs, made fire-resistant by treating the tar with unslaked lime and surfacing these shingles with sand. W. A. Lampadius was the man who promoted the use of such "tarred board" shingles in Germany [2]. In the meantime Neufchatel asphalt had been found to be excellent material for impregnating porous paper for use as tarpaulins, packing paper, imitation-cloth, etc. and manufacture was started at Geneva [3].

A true revival, however, did not take place until about 1840. Good roofing paper was now prepared by immersion in hot tar. Wood-tar being too expensive, coal-tar was now generally used, which could be obtained cheaply as a byproduct of the gas factories. Rolls of felt-paper 0.75 to 1.00 m. wide were suspended above a bath and the sheets were drawn through the tar, excess tar being expressed by two rollers through which the impregnated paper ran before it was hung on drying stocks. A better application was the successful

---

[1] Anon., Einige historische Notizen über Dachpappen (Teer und Bitumen, vol. 34, 1936, pp. 332 ff.).
[2] J. f. techn. u. ökon. Chemie vol. 6, 1829, pp. 377 ff.
[3] Kunst- und Gewerbeblatt 1820, p. 589.

use of shingles cut from such paper. Instead of nailing the shingles directly on to the planking, they were now nailed to laths laid across this planking which avoided the frequent rupture and distortion of the shingles as the woodwork expanded or shrank. Using strips of impregnated felt instead of shingles also avoided too many contact points by nailing and thus the roofing had a longer useful life. Asphaltic residues were also used from time to time to manufacture such paper and "double-felt roofs, gravelled roofs and wood-cement roofs" were now the most favoured types of roofing and were cheaper than the older roofing tiles and much lighter.

Similar developments took place in the United States in those years. Rev. Samuel M. Warren reported that in 1844-1845 roofs were first laid in Newark, N. J., consisting of square sheets of ship's sheathing paper treated with a mixture of pine tar and pine pitch, and surfaced with sand. In 1847 coal tar was used by Cyrus M. Warren as a substitute for the pine tar, to soften the pine pitch, and employed as a saturant for the paper. Fine gravel was next used to substitute the sand. The square sheets were dipped into the melted mixture by hand, sheet by sheet, and then the excess was pressed out. The next step consisted in running the paper or felt in rolls through continuously operating saturators designed to saturate with tar. Finally, in Buffalo, N. Y., coal tar was distilled down to a roofing pitch, which was used to replace the more expensive mixture of pine pitch and coal tar. The foregoing constituted the origin of the "tar-and-gravel roof", so extensively used today.

Perutz [1] informs us that the impregnation vessels measured 6 to 10 feet across and that they were 1 to 1½ feet deep. He advised not to impregnate first with a mixture of bitumen and minerals as some manufacturers were want to do. The first bath should contain a fairly soft bitumen, well distilled and containing no water or light fractions. The second bath should hold a harder bitumen, if required mixed with minerals. A good roofing felt should not melt in the sun, nor should bitumen drip from the roof. It should be covered with so much sand and ashes that it charred only in case of fire and that it would never burn with a flame. In his days apart from the Seyssel and Neufchatel mastic, bitumens from Trinidad, Cuba, Canada (100 miles east of Detroit), Virginia (Cairo, 30 miles east Parkersburg), Albertite from New Brunswick and asphalt from Coxitambo (Peru) were sold on the European market in order to cope with the demand which far exceeded the available amounts of coal-tar pitch.

The bituminous felt was also put to another use, the manufacture of "bituminized paper pipes" which for several decades held their place against the

[1] H. Perutz, Die Industrie der Mineralöle (C. Gerold, Wien, 1868, pp. 175-176).

expensive cast-iron pipes then on the market. Their manufacture is described in the following words in a contemporary periodical [1]:

"These pipes are made on a cylinder of which the diameter is the required bore of the pipe; the paper is rolled in widths equal to the required length of the pipe, and passes through molten bitumen, until the required thickness of pipe is attained. Acting in connexion with this, another cylinder revolves against the first cylinder bearing the bituminized paper, the latter being thus subjected to a very great pressure, which causes a regular distribution of bitumen. When the pipe is taken off the cylinder, the inner surface is coated with an indisoluble water-tight composition, and the outer surface is coated with bitumen mixed with sand. The compressed paper, without the molten bitumen forms one-third of thickness of the pipe. The great advantage of these pipes consists in their resistance to internal and external pressure. Their power of resistance is sufficient to withstand a pressure of more than 240 pounds to the square inch, or 15 atmospheres, equal to more than the pressure of a column of water 500 feet in height. In Hanover and Paris, where experiments were made with these pipes in reference to their resisting power, they withstood a pressure at the former city of more than 24 atmospheres, equal to about a column of water 800 feet in height, and in the latter city they withstood a pressure of more than 500 pounds on the square inch, equal to about a column of water 1000 feet in height. In making these trials not only one pipe, but several joined together were tested. As above stated the resisting power of the pipes can be considerably augmented by adding to the thickness. The manufacturers undertake to furnish pipes to resist a pressure of 600 pounds on the square inch. Judging from the experiments made by Mr. Samuel Hughes, Civil Engineer of London, and others, the resistance to external pressure of the bituminized pipes is so great that they can with perfect safety be laid under any depth of causeway, and bear equal or unequal pressure.

Durability. — The known property of asphalt to resist atmospheric changes, renders it most valuable for pavements and roofing, whilst when applied to pipes, exposed as a rule to the humidity of the soil only, it must augment their durability to an almost unlimited degree. Iron pipes are nearly always coated with asphalt to increase their durability and protect them, but the paper pipes, being composed in a great measure, throughout their substance of this material, must have a great advantage over them. More than ten years experience shows, besides, that paper pipes that were laid in Paris in the year 1851, when taken up were found to be in the same state as when laid down. As these pipes do not suffer either from concussion or frost, and are not liable to leakage or oxidation, (the causes of destruction to metal pipes), their durability must needs be almost unlimited.

Impermeability. — By the above described mode of manufacturing, a perfect homogeneous substance is attained, and the density of the bituminized material is greater than that of all other kinds pipes. Bitumen a Bad Conductor. — Bitumen being a bad conductor protects the water in these pipes from excessive cold in winter, and in summer from heat; this same peculiarity farther leaves the length of the pipes unaltered by the change of temperature; this is not the case with metal pipes.

[1] Anon., Bituminized Paper Pipes and Roof Slatings (J. Franklin Institute vol. XLIX, 1865, pp. 210-212).

The inoxidability of the bituminized paper pipes is another circumstance greatly in favor of their use as a substitute for iron tubes. Such inconveniences as destruction or obstruction by oxide, noxious chemical alterations of water, as in the case of leaden water pipes, impregnation of the water with oxide of iron, so very injurious for many purposes cannot take place with the use of tubes made from bituminized paper, neither is there the most trifling impurity imparted to the water passing through them.

Bitumen unaffected by Acids and Alkalies. — The bitumen pipes are not affected by acids or alkalies, and remain perfectly unimpaired when water, impregnated with acids or alkalies, is led through them. Metal pipes, on the contrary, are subject to rapid decomposition when placed in sulphuric acid, chalk soil, or copperas in solution, frequently found in coal-pits and mines. Non-conductor of Electricity. — Through the peculiarity of bitumen not being acted upon by acids, and being a non-conductor of electricity, the bituminized pipes are not liable to destruction and stoppage, as iron pipes are in which galvanic streams are frequently produced, causing leakage and oxidization. In virtue of this peculiarity, asphalt pipes will be used with advantage for underground telegraph wires, through tunnels, and under bridges.

Elasticity. — The elasticity of these pipes protects the same from bursting by external shocks or concussions, which causes most of the other kinds of pipes to break, the capability, also of extension is greater than the expansion caused by frozen water, hence bursting is avoided, an accident to which all other pipes are liable.

Specific Gravity. — The specific gravity of these pipes, compared to iron pipes, is as 1 to 5, which circumstance, as it facilitates transport and adjustment, considerably reduces the cost.

Cheapness. — In Great Britain, where all kinds of metal pipes are undoubtedly manufactured cheaper than anywhere else, the bitumen pipes, when laid, are still only one-fourth the cost of lead, and about one-half the cost of iron. On account of their great lightness, the rate for transport is comparatively very small—say only one-third of the cost of transport for iron pipes. In conclusion, it may be mentioned, that all workings, repairs, etc., of bituminized pipes are much easier executed and at less expense than is the case with other pipes. The following bituminized pipes are manufactured in seven feet lengths:

A. Water pipes to withstand the pressure of 15 atmospheres, either with bituminized couplings or with iron flanges.
B. Gas pipes (lined with metal to resist the chemical influence of gas) with iron flanges.
C. Air pipes for mining purposes, of lighter construction, and connected by bituminized couplings."

From the evidence given it is clear that by 1880 the European refiners were well on their way to the manufacture of good fuel oils, lubricants and bitumens from the crude oils at their disposal and that the variety of their crudes and the complexity of the European markets for such products had not deterred them from making strenuous efforts to succeed. In many cases their efforts contributed essential methods and data for the future rise of the petroleum industry. Drake's European contemporaries have not fought their battle in vain, they established modern petroleum refining in all its variety.

INDEX

Acid wash 114, 115, 116, 125, 152, 153, 154, 155, 157
acidity (luboils) 177, 180
Aegineta, Paulus 23
Aeneas the Tactician 72, 73
agates 27
agitator 116, 128
Agricola, Georgius 1, 12, 26, 37, 39, 40, 41, 46, 51
Agrigento (Sicily) 13, 14, 19, 27, 34, 41
air pollution 126
Albertite 190
Alfreton 61, 62
Alsace 12, 92
amber 13, 14, 40, 41, 46, 47, 49, 52
Ammianus Marcellinus 41, 73, 78
ampeline 149
ampelitis 28, 29, 42, 44
A.P.I. (American Petroleum Institute) 68
Apollonia (Epirus) 21, 22, 24, 35, 40, 42, 51
Arcet, d' 118
Argand, F. P. A. 108, 109, 110, 111
aromatics 58, 66, 106
asphalt, see bitumen, asphaltic
asphaltic compounds 155, 178
asphaltos 3, 4, 11, 19, 24, 33, 42, 48, 56
Astagene 21
astatki 161
Atchin 47
Audoin, P. 64, 166, 167
Aurifaber 46
autoclave process 146
Auvergne 50, 57, 62
Avicenna 22
Aydon, H. 167, 168, 169

Babylonia 34, 40, 41, 49
Baku 62, 64, 106
balloon 108, 109, 110

Barbados 35, 47, 62
baroud 89, 90
Bastenne Bitumen Co. 184, 185, 186
Baumé, A. 51, 69
Baumhauer, E. H. van 65
beeswax 144, 145
Beilstein, H. 66
Belmontine 150, 152
Belon, Pierre, 1, 2, 13, 16, 23, 25, 29, 31, 93
benzine 66
benzole 66
benzoline 66
Berzelius J. J. 54
Biddle, J. E. 163
Binns, Thomas 145
bitumen, 3, 4, 8, 20, 28, 33, 34, 40, 42, 43, 44, 46, 49, 51, 56, 183
applications 183
asphaltic 24, 94, 183, 190
distillation 30
glue of, see naphtha
Judean, see Dead Sea Asphalt
liquid, see petroleum
origin of 25, 37
pipe coating 191, 192
production 8, 40
road 183, 184
roofing 184, 187, 188, 190
Blanckaert, Stephen 32, 34
bleaching 153
bloom (of distillates) 120
Boerhaave, Herman 51
Boghead coal 113
bomb, incendiary 85, 86, 87
bone black 156
Boryslaw (Galicia) 93, 94, 96, 97, 98
Boulduc 50
Boussingault, J. B. 63

Brasavola, 22, 25
Brez, Sieur de 145
Brocardus 20
Brunswick 12, 32, 40
Burma 50, 56, 61, 151, 152, 153, 154
Burner, 161
   air pulverizing 167
   Astrakhan 161
   atomizing 165
   Aydon-Selwyn 169
   Guyot 167
   Holden 169
   Linton-Shaw 164
   nozzle-type 161
   pressure atomizing 169
   slot-type 161
   steam-pulverizing 162, 163, 169
   tube-type 161
   Venturi 161

Cahours, A. 64
Calao (Peru) 49
Cambacères 145
Camphene 141, 142
camphor 14, 41, 42, 48
camphoride 147
Campina (Roumania) 101
Canada 136, 190
candle 143, 144, 145, 146, 151, 152
   beeswax 144, 145
   making 145
   paraffine wax 147, 150, 152
   stearin 146
   tallow 144, 145, 146
carabé 29, 46, 52
carbo 43
carbon, animal 156
   tetrad 66
   vegetable 156
carburetting 160
Cardanus, G. 46
Carpathian Mts. 42
Carthage 12, 40
cart grease 94, 97
Cassius Dio 76, 77
casting candles 145
Celisi, Mt. 4, 5

ceresine 97, 157
charcoal burning 18
cheirosiphona 87
chemistry, organic 54
Chesebrough 157
Chevreul, M. E. 46
Cissia (Susa) 9
Clermont 62
coal 37, 41, 43, 44, 52, 53, 161, 162, 164, 165, 166, 169
   cannel- 91
Coffey, J. A. 118, 123
Cogniet, C. 69
Constantine Porphyrogenitus 80, 82
copal 49
Cordus, Valerius 1, 14, 31, 32
Coulomb, Ch. A. 172, 173, 174
cracking 58, 106, 117, 137
crude oil, see petroleum
Cuba 190
Cymogen 122

Dead Sea 13, 14, 20, 21, 22, 24, 27, 33, 44, 49
Dead Sea Asphalt 24, 25, 33, 35, 37, 40, 44, 57
Deville, H. E. Saint-Claire 64, 65, 166, 167
Dieudonné 65, 166, 167
Diodorus 13, 73
Dioscorides 3, 16, 17, 18, 20, 22, 26, 29, 36, 42, 44, 45, 93
distillation 30, 58, 61, 113, 115, 117, 127, 151, 152
   continuous 118, 119, 120, 122, 123, 127
   dry 148, 149
   steam 153
   tests 138
   vacuum 117, 118, 124
Dolfuss, Ch. 179, 180
Dordtsche Petroleum Co. 65
Drescher, J. 188, 189
drilling, Canadian cable 97, 98, 105
Drohobycz (Galicia) 93, 97
Dumas, J. B. 149
Dutch East India Co. 47

INDEX

Ecbatana (Iran) 4, 12, 40
elaterite 57
electricity 134, 135
engine, internal combustion 160
Evax Maurus 45

Falarica 73, 78
Faraday, M. 133, 158
Farez 177, 178
Faxe, Arvid 186, 187
Fellin 21, 24
Figuier 156
fistula 36
fire, arrows 72, 73, 76, 78
  brands 70, 71
  chemical 75
  Greek 62, 70, 79, 80, 81, 83, 84, 86
  liquid 84, 85, 86, 89, 90
  prepared 81
  self-igniting 76, 84, 85
  self-propelling 84
  ships 71
  wild- 84
flamethrower 74, 75
flash point (fire point) 136, 137, 138, 139, 177
food preservation 158
fossil carbon, see coal
fractionation 115, 125
Frankland 134
friction 172, 173, 174, 175
  dry 174
  immediate 172, 173
  mediate 172, 173
  sliding 172
  wet 174
Fuchs, L. 21, 25, 26
fuel 161, 162, 164, 165, 166, 167
  locomotives 65, 162, 163, 165, 167
  motor 160, 166
  ships 64, 67, 162, 164, 165, 166, 167
Fuhst, H. 122, 123
fuligo, see lamp black

Gabian (France) 14, 35, 62
gagates 13, 21, 25, 26, 27, 28, 33, 34, 42, 43, 46

Gale, Thomas 141, 159, 160, 163, 164, 170, 171
Galen 13, 14, 18, 19, 20, 21, 23, 24, 26, 28, 43, 44
Galicia 24, 54, 62, 92, 93, 94, 97, 98, 136
gas, coal- 134, 145
gasoline 151
gasoline process (wax refining) 154, 155
Gay Lussac, L. J. 149
Gesner, Abraham 115
glass, lamp- 110, 111
Globoil 176
Görsdorf (Alsace) 12
graphite 180
gravity, specific 68, 69
grease, lubricating 169
Greek fire, see fire, Greek
grenade 87
gum, see resinous compounds
gunpowder 83, 84, 86, 87, 89, 90

Hannover 40, 137
Hatchetite 150
Hatchett, Ch. 57, 150
Hecker, Joseph 93, 94
heptane 67
Herodotus 9, 13, 70
Herzberg 187, 188
Hirn, G. A. 173, 174
Hochstetter 156, 157
Homer 70
hydrometer 58

Illuminants 130, 131, 133
  candle power 130
India 41
Indonesia 65
Isherwood, Comm. 165
isoprene 57

Jet 34, 42, 43, 45, 48, 49, 52
Julius Africanus 84

Kallinikos 80
Kaukasine 176
kenderbal 99
kerosine 93, 114, 116, 117, 130, 134, 153
kipiaczka 93

klysteros 79, 86
knipas 43
Körting 161, 169
Kurbatow 66

Lacka 93
lamp black (soot) 17
lamp oil (photogen, paraffin) 11, 12, 62, 93, 94, 97, 102, 112, 113, 114, 139, 141, 142
lamp, oil- 108, 109, 110, 111, 112, 115, 116
Lampadius, W. A. 189
Lange A. B. 109, 110, 111
Laurent, A. 149
Lavoisier, A. L. de 54
Lenz, O. 161
Leo VI (emperor) 81
Libavius, A. 14, 46, 47
lighting (oil-) 94, 102
lignite 91
Likhatchoff 64
lithantrace 46
Livy 71, 72, 73
Lobsann (Alsace) 57
Lôme, Dupuy de 65, 166, 167
Lowitz, K. 156
lubricants 94, 97, 114, 116, 125, 157, 170, 171, 172, 174, 175, 176
  testing 179
Lucullus 48
Lugo 120
Lukasiewicz, Ignacy 94
Lycia 26, 39, 43
lyncurium, see amber

Macquer 51
malleoli 73
maltha 39, 42, 46, 49, 56, 77, 87
mastic, bituminous 183, 184, 185
Matthiolus 3, 16, 19, 24, 29
mazout 161
Medea 12, 41, 48
Media 12, 21, 25, 77, 78, 86
Memphitic stone 28
Merenda 25, 26
Mesuë 26
Miano (Italy) 4, 10, 91, 92
mining laws 97

minyak tanah 47, 48
Modena (Italy) 4, 5, 7, 13, 16, 22, 25, 34, 35, 40, 49, 50, 56, 91, 92
moderateur lamp 111
montejus 124
Montgolfier brothers 109, 110
Morin, A. 172, 173
mortar, bituminous 43, 48 (see also mastic)
Müller, H. 61, 150, 152, 153
mûmiâ 3, 19, 22, 23, 29, 33, 34
  aegyptiaca 25, 34
  artificial 22, 23
  natural 22, 23, 34
  primaria 25

Naphtha 3, 4, 6, 8, 10, 20, 21, 32, 33, 34, 42, 46, 48, 49, 52, 53, 56, 57, 61, 62, 67, 68, 77, 86, 87, 89, 90, 115, 116, 125, 133, 136, 137, 149, 151
  analysis 58, 59
  adulteration 11
  black 4, 6, 7, 8, 10, 34
  Boghead 57
  bone 57, 59, 60
  coal 60, 66, 67
  inflammability 9, 11, 12, 22, 43
  nature, see petroleum
  production 5, 57
  red 3, 4, 5, 6, 10
  refining 58, 61, 62
  shale 67, 68
  white 4, 10, 34
naphthenes 66
naphtometer 139
Napoleon III 64, 166
Nasmyth, H. 179
Neufchâtel (Switserland) 49, 62, 181, 190
neutrality (luboils) 177
Nicander 28, 45, 51
nitric acid 152

Obsidian 12, 34, 42, 44, 45
oil, castor 180
  coal 112, 132, 135, 155
  colza 11, 112, 180
  cotton-seed 180
  earth- see petroleum

fatty 112, 130, 131, 144, 169
incendiary 78, 79
lard 181
neats-foot 180
olive 112, 176, 180, 181
patent 111, 112
price 10, 105
rapeseed 176, 180, 181
sesame 112, 180
sperm 176, 180
strata 8
tallow 130, 170, 180, 181
trade 11, 97, 105
transport 105, 106, 162, 165
whale 180
Olearius 50
oleine 146
Oleonaphta 176
oleum benedictum 78
oleum incendiarum 77, 78
olio de sasso 32
oly van aerde, see minyak tanah
Oribasius 19
Oxus river 12
ozokerite 150, 157

Păcura 101
paint, bituminous 41
paraffin, see lamp oil
paraffin oil 113
paraffins 66, 67, 149, 150
Pechelbronn 57, 91, 173, 176
Pegir (Atchin) 48
Pelouze, J. 64
Pennsylvania 91, 136, 142, 143, 159, 160, 169, 176, 178
Persia (Susiana) 25, 27, 33, 34, 42, 47, 50, 55, 61
Perutz, H. 120, 122, 123, 156, 190
petroleum 3, 6, 32, 33, 34, 36, 42, 46, 48, 50, 51, 56, 61, 62, 93, 125, 139
  medical value 159, 160
pharmacites 28, 44, 46
Philostratos 72, 75, 76
Phocion 43
photogen, see lamp oil
photometer 130

Pietro, Mt. S. 6
pipes, bituminized paper 191, 192
pissasphaltos 3, 19, 20, 21, 24, 25, 34, 45, 42, 46, 48, 49, 50, 51, 93
pitch, Bruttian 19
  coal-tar 17, 18, 25, 34, 42, 49, 76, 77, 189, 190
  Jew's 3, 4, 19, 20, 25, 33, 48, 49
  liquid 17
  oil 17
Pitchford (Shropshire) 49, 57
Pliny 9, 12, 19, 21, 22, 24, 27, 28, 37, 39, 41, 42, 43, 44, 45, 76, 77
Ploesti (Roumania) 102
Plutarch 4, 73
Poland 63
Polybios 75
Porta, G. B. della 48
Poseidonius 28, 44
Pratt's Astral Works (USA) 126
Price and Co. (London) 146, 151
Procopius 71, 77
Puebla 64, 65, 166, 167
pyr, hygron 82, 83
  Medikon 83
  thalassion 80, 83
pyrphoros 75

Quinquet, A. A. 109, 110, 111

Rangoon 47, 61, 63, 92, 136
Ragosine 176
Ragusa (Sicily) 50
Redwood, Boverton 69
refineries 97, 98, 102, 103, 104, 120
refining (petroleum products) 30, 42, 58, 60, 114, 116, 125
Reichenbach, Carl von 133, 147, 148, 149, 150, 152
resin 29, 76, 82, 83, 130
resinous compounds 176, 177, 180
Revina 5, 6
Reynolds, O. 174, 175, 176
Rhadinace 9
Rhigolene 122
rock oil 34, 51, 139, 141, 163, 170
rockasphalt 183
roofing (felts, shingles) 186, 187, 188, 189

ropa 93, 94
rosin 18
Roumania 11, 14, 61, 99, 101, 102
Royal Dutch 65
Rubinoil 176
Rue, Warren de la 61, 116, 151, 152, 153, 154
Rumford, Benj. Thompson Count 130
Russia 62, 63, 105

Salsa 4, 7
salt, manufacture 8
salt domes 101
Samosata 40, 42, 48
Samothrace 42, 43, 45
Sassuolo (Italy) 5
Scheererite 149
Schorlemmer, C. 64
Scotland 45
Seefeld (Tyrol) 16, 21, 26, 29
Selligue, J. 116
Seneca oil 159
Serapion 4, 22, 23, 42
shafts, hand-dug 94, 96, 99, 101
shale, bituminous 41, 113, 117
 oil- 62, 149, 176
Sherwoodole 151
shingle (roofing) 186, 187
Sicilian oil 41
Silliman, B. 63, 64, 115
sinumbra lamp 111
sludge 60, 114
smelting (with fuel oil) 169
snuffing candles 145
soda wash 114, 125
sodium plumbate 124
Soli (Cilicia) 12, 41
Solinus 45
Spakowski 161, 162
sperm oil, see oil, sperm
spermaceti 131
spinos 44
spirit, petroleum 66, 67
stability (luboils) 177
Staroil 176
stearin 146
stearopten 147
Stephanus 43

stone oil, see petroleum
storage, petroleum 139
Strabo 12, 22, 41, 43
strepta 81, 85, 86, 87
sublimation 31
succin 29, 52
sulphur 37, 49, 72, 73, 76, 77, 82, 83, 89, 90
sulphuric acid, see acid wash
syphons 80, 84, 85, 86, 87, 89
syringes, see syphons

Tallow 144, 145, 146, 180
tar, Barbados 47
 coal- 61, 63, 91
 Rangoon 47, 150, 151, 153, 176
 wood- 17, 63, 91, 186, 187, 188, 189
Tegernsee (Bavaria) 12, 16, 40, 62, 150
Theophrastus 18, 43, 44
Theopompus 43
Thracian stone 27, 28, 33, 42, 43, 45, 52
Thucydides 70, 71, 74
Thurston, H. 174, 175
Tdirschenblut 19, 21, 25
Titusville (Pa.) 63
Topazoil 176
topping crude oil 122
Torbanite 113
tow 71, 72, 73, 83
Tower, Beauchamp 174, 175, 176
Tralles 12
Travers (Switserland) 181, 184, 185, 186
trebouchet 88
Trinidad 56, 57, 62, 63, 182, 185, 190
Twitchell 146

Ujhely 97, 157
unctuosity, see viscosity
Ure, A. 55, 56

Valona 21
Valvoline 176
vaseline 157
Vegetius 78
Venango Country (Pa.) 62
vinegar 72
viscometer 180
viscosity 177, 179, 180

Vohl 136
Vossius, Isaac 83
Vulcanoil 176

Wagemann 116, 117, 176
Warren, S. M. 190
water, separation 5
wax, mineral 21, 24, 34, 96, 97, 149, 150
   mountain 96, 150
   paraffin 61, 62, 112, 122, 147, 149, 150, 151, 154, 155, 156, 157, 158, 180
   pressed 154, 155
   slack 154, 155
wick, candle 145
   lamp 108, 110, 111

Wiesmann & Co 150, 176
Wietze (Hanover) 91, 92
Williams, C. G. 57, 60, 61, 62, 63
Wilson, G. 146, 151
wisk 99

Xenocrates 45
Xenophon 71

Young, James 113, 116, 132, 150, 153, 154

Zacynthus (Zante) 19, 42, 49, 62
Zibio (Gibio) (Italy), Mt. 4, 5, 7, 40, 62

LIBRARY OF DAVIDSON COLLEGE

Books on regular loan may be checked out for **two weeks.** Books must be presented at the Circulation Desk in order to be renewed.

A fine is charged after date due.

Special books are subject to special regulations at the discretion of the library staff.